CONSTRUCTIVE COMMUNICATION
Skills for the building industry

Richard Ellis
Consultant

ARNOLD

A member of the Hodder Headline Group
LONDON • SYDNEY • AUCKLAND

First published in Great Britain in 1999 by
Arnold, a member of the Hodder Headline Group,
338 Euston Road, London NW1 3BH

http://www.arnoldpublishers.com

British Library Cataloguing in Publication Data
A catalogue record for this book is available from the British Library

ISBN 0 340 72007 7

1 2 3 4 5 6 7 8 9 10

Publisher: Eliane Wigzell
Production Editor: Julie Delf
Production Controller: Priya Gohil
Illustrator: Thomas Laird

Typeset in 11/13 pt Palatino by Phoenix Photosetting, Chatham, Kent
Printed and bound in Great Britain by
J W Arrowsmith Ltd, Bristol

What do you think about this book? Or any other Arnold title?
Please send your comments to feedback.arnold@hodder.co.uk

Contents

Acknowledgements

This book owes a great deal to the close involvement during its preparation of a number of individuals who kindly gave of their time to assist the author with their expert views.

Ian Clark, Senior Lecturer, Department of Building, Edinburgh's Telford College

Mark Glancy, Study Centre Manager, Edinburgh's Telford College

Steve Willis, Lecturer, Crawley College, Crawley, Surrey

Martin Hassett, Lecturer in Wood Trades, St Helens College, Merseyside

John Laws, Head of Wood Trades, Cambridge Regional College, Cambridge

W C Ryan, Lecturer in Construction and the Built Environment, College of North West London, London

Thanks also to George Grams, Ronnie Turnbull, Eliane Wigzell at Arnold and Thomas Laird, the illustrator.

Special thanks to Grace for all her assistance and patience.

Introduction

Developing your communication skills will help you in all kinds of ways whether you are at college, working for a company or working for yourself. You can have all the right skills as a joiner, plasterer, mason, builder, plumber, etc. but if you are not able to:

- take messages effectively;
- write a clear letter or invoice to a customer;
- phone though orders in a confident, business-like manner;
- deal with complaints without blowing your top;
- complete a health and safety checklist so that no details are lost;

then you will be that much less useful to those you work with, your customers and your suppliers. If you decide to set up your own business then being effective in communication with customers and colleagues will be essential to any success. Such skills may determine whether you stay in business or not.

The aims of this book are to:

- explain some techniques that will help you with communication;
- increase awareness of the importance of communication at work;
- provide opportunities for practising some of the key skills.

You may be taking a course at college, studying for a NVQ or a SNVQ. You may have taken your courses, passed all your exams, gained your certificates and now be working for a company. You may have started working for yourself. You may have this vision of you the boss, with your vans going round your town or city carrying your name to hundreds of houses! You may actually have a van and look forward to the day when you'll have a whole fleet!

Whatever your situation, age or experience in the construction industry we hope that this book will be of help to you. It will cover a good deal of what you will need to know about communication, especially on-site and with the customer. We will show you ways of improving your letter writing skills, handling phone calls and

taking messages, as well as many other aspects of communication such as:

• applying for jobs
• filling up forms
• being confident in meetings
• presenting ideas to customers and suppliers, etc.

A book can only do so much. We hope you will take the ideas and try them out for yourself.

We are confident that by reading about these ideas and trying out the activities you will be in a better position to notice good and not so good communications. Through this observation you can then learn

what to adopt and what to avoid when it comes to your own practical communication.

It's seldom a good idea to copy someone else. What you can do, and what we would certainly recommend, is adapt others' communication skills and approaches and select those aspects which best fit your particular style and the situations you find yourself in.

The comparison here would be rather like looking at someone else's clothes and admiring them; you wouldn't necessarily go out and buy the same but you'd take the idea – the style – and find something similar but not identical for yourself to wear. You'd choose something appropriate to you.

Communication skills

Communication skills are similar to other skills. You didn't learn to ride a bicycle straight off, remember you had a go, fell off a few times, picked yourself up again and had another go. You probably had someone alongside who could actually ride who guided you and gave advice.

We'd like to think that the ideas and opportunities for practice outlined in this book will help you 'ride' better and go further. Many of the exercises are designed for you as an individual to carry out; others are better done as group work; these you'll find at the end of each chapter.

It is essential though for you to do some practice whether on your own or as part of a group in class. We can suggest ways of communicating more effectively but you will need to work through the activities and reflect on how you could adapt them to your work. Do try them. Try to work out your solution before you turn the page or read further ahead.

We also provide a number of checklists for you to think about and complete. They are not designed to be the final say, a complete listing of all factors, but as a way of encouraging you to think about your communication.

We've also added a number of examples of different communication taken from various companies in construction, plumbing, joinery etc. We encourage you to read these carefully and be critical of them. Think as you read them of ways in which they could be developed and improved.

However, before we get down to details of letter writing, dealing with customers' enquiries and complaints (you'll be sure to get some of these) we need to think about some core questions relating to all our communication whether by fax, phone, mail, chat over a coffee or video link! (Yes that'll be with us all soon.)

Activity

Spend a couple of minutes jotting down what you think should be the key questions to ask of any communication. This could be a letter, report, telephone call or interview. Then compare your list with ours.

Ten questions to ask about any communication

(1) Is it necessary?

There's so much communication flying about these days, so much that gets thrown straight in the bin, that we should ask ourselves at the very start: Is this letter, fax, notice, e-mail, meeting, telephone call necessary? We're not saying that the occasional social/telephone call/visit isn't a good thing but people are more and more time-conscious and will resent their time being wasted.

So make sure that any communication is necessary and so less likely to end up in the waste paper basket or being ignored. These other questions are designed to help you towards that goal.

(2) Is it targeted?

You know just how disappointing and frustrating it is when you tear open that interesting looking letter only to find that it has been misdirected and doesn't concern you at all. Or perhaps even worse when you open the envelope to find that some firm is refunding your money – an overpaid bill from way back – only to find after you've read the letter several times that the refund is not for you at all but for your neighbour. So our first consideration when it comes to successful communication is – aim it at the right person/s. Make sure you have his/her name and that you spell it properly (Whyte not White) not to mention the right address (121 Closewynds Avenue not 112 Crosswinds Avenue).

The communication must hit the target if it is to have real impact.

It is essential for you to know who your customers are and what their needs are, not to mention important details like their names, designations (the jobs they do) their business and home address, telephone and fax numbers. See the use of record cards on pages 100–101.

Whatever you do avoid just firing off a letter to a firm. Try if at all possible to find out the name and job title of the 'target'. If you want to persuade them to use your firm, try your services or take up your offers then address the letter to Mr Harry Smith not The Manager, The Owner, The Site Foreman, etc. Sometimes of course it is just not possible to find out someone's name. In that case try and make sure that you have the correct title. People can get quite annoyed if they're called Engineer rather than Senior Site Engineer or Chief Assistant rather than Assistant Chief!

Good businesses are built on such small details.

Finding your key target customers and suppliers and keeping them in your sights is part of that detail. Neglect this at your peril. Successful communication in business is about how you build up and maintain a good relationship with customers, suppliers, health and safety inspectors and work mates. Getting one contract from a customer is fine but getting three or four jobs every year for five years from the same customer is what you need to be aiming at.

(3) Is it timely?

Does your communication come at the right moment? If you have been asked to estimate for a job then there is no point in sending your letter with the quote three weeks later – your potential customer will probably have found someone else. Strike while the iron's hot! Don't delay. A rapid response will add a positive image to you and your organisation. Any delay might be thought of as a lack of enthusiasm, that you couldn't be bothered or that you don't have very efficient systems in your office.

Don't put off answering that letter or request for information. These small seedlings may grow into a substantial part of your company's business. Often the best customers start off by buying just a very small service from you, a nibble, before they go for the whole chunk; they like to taste first before committing themselves. Don't you? So don't neglect the small orders – remember acorns grow into oaks.

Don't miss out on those opportunities which often present themselves on buses and trains. Those chance encounters can lead to business. You happen to sit next to someone on the train and he gets talking about where he works and the fact that there's so much building needing to be done, never mind all that repainting. This is when you offer your card (which you should always have with you) and just maybe that chance encounter will actually lead to business in the future. Close encounters of the lucky kind!

Being at the right place at the right time is often just luck. Having a business card ready so that you can sell your skills is preparing to

succeed. Having a big, clear and clean noticeboard up at the site where you're working is preparing for your next customer!

(4) Is it in the right language?

No we're not talking about you writing your leaflets in Chinese or making calls in Italian (though if you can speak another language then do take advantage of it). We're referring to English. But which English? The construction industry, like others, has its own special language.

Activity

Here are some examples. Do you know which section of the construction industry they come from?

(A) Perform all mitres, stopped, returned, splayed and mitred ends, moulded, rebated, sunk, weathered and throated work.
(B) Quirk in internal angle to localise all differential cracking.
(C) Socket outlets with flush pattern ivory plastic switched (unswitched).
(D) All holes provided with bolts sufficiently long for the purpose with washers, taper pattern being used.
(E) To be jointed with approved compression fittings (capillary fittings, silver brazing or bronze welding).
(F) Solid hard and level inserts of the same veneer may be provided to repair the face but no end joints are permissible.
(G) The face shall be well cleaned down to remove all cement lattice, well wetted and covered with a thin layer of equal portions of cement and sand.

These special languages for joiners, bricklayers, electricians, plumbers are fine when spoken amongst themselves; the problems come where they are used with the public and with those outside the industry.

We need to put our language into theirs – plain English e.g.

We need some GRP round this MDF.

Plain English: 'We need some glass reinforced plastic round this medium density fibreboard'.

This has got shake.

Plain English. 'I'm afraid this piece of timber won't do – look at these defects'.
It's not only the 'foreign' words like shake, closing stile, expanded metal lathing, stanchion bases, etc. which give problems and will need to be translated into ordinary English. It's words which sound familiar such as eaves, set, blooming, which have a very different meaning in the industry from that used by ordinary people. Looks like it's blooming has nothing to do with growing roses but translated means, *'Looks like you've lost some of the gloss finish – probably damp getting in'*.
Think of the times you've sat and scratched your head when trying

to understand some letter from the garage, building society, solicitor, etc. puzzling over what they mean in their special language. *'If only'*, you say, *'they could put this into plain English'*. Well, let us be sensitive and aware of the need to carry out this translation and explanation for our readers and listeners.

(5) Is it clear?

We'll cover this in some detail in this book, but we're sure you can realise the effect on your listener/reader if you are not clear what you want e.g.

> Dear Mrs Brown
> It'll be a couple of days – perhaps tomorrow – before we get the painting done; it can't be done before that because the bloke can't get here – he's busy somewhere else – he's got most of the stuff – but I'll give him a phone to see if he could speed up a little – Is that OK?

Well it might be OK to you but pity the poor customer! This is a muddle. *Because the bloke can't get here?* What will this mean to Mrs Brown and what message does the phrase: *he's got most of the stuff* imply? Would you in all honesty want to buy a second hand car, or have your house decorated by a person who was so confused in what he was trying to say?

A muddled communication often gives the impression that our actual work won't be that great. Sloppy communication often implies that a sloppy job will be done. The customer may well think, if they can't get that detail right then can I trust them over that rewiring, plastering, or chimney repair? Wouldn't you feel that way if it was your car being serviced, your video repaired, your holiday being booked and a small but probably significant mistake was made? Your faith in the firm's abilities would probably be somewhat dented.

(6) Is it accurate?

We've already mentioned the importance of getting names right. Being accurate is vital when we are setting out dates, times, measures, quantities, costs, etc. Some very expensive mistakes have been caused by leaving off the odd nought – £10.51 not £100.51. How many jobs have been lost simply because someone's left out the odd digit from a telephone number, or thought a 1 was a 7, and so the potential customer who phoned in with a request couldn't then be contacted. It would be typical bad luck (Murphy's or Sod's Law) if that was the customer who was just thinking of having a large new conservatory and was going to put the whole job in your lap!

Thinking about mistakes, you may have had the unpleasant experience of your bank giving you the wrong information as regards your account. You go to the hole in the wall, enter your bank card and punch in your pin number and then stand back amazed at either how little you've got left or how much. Some have been known to have heart attacks on the spot. Unfortunately even if on the slip it says that your account has £100 010 in it, while in fact you've only got £10, you can't go out and spend, spend, spend; the law states that any reasonable person should know that a mistake has been made – unless that is you are in the habit of having £100 000 regularly placed in your bank account!

They happen so easily, these little mistakes which can cause enormous problems. Apparently no-one could find the binoculars for the lookouts on the Titanic, so when they finally saw the iceberg it was too late to turn the ship which was steaming ahead at full speed. It was a simple mistake which had enormous consequences! **The lesson is take care of the details**.

(7) Is it short and to the point?

Because your reader or listener is usually busy he or she will not want to plough through lengthy letters, reports or proposals, nor will your rambling phone call be welcome. It is important to ensure that all our communication at work, with customers, suppliers and colleagues, is as concise as possible. However, we should take care that it is not too short; then it can easily become abrupt and rude. The following is nice and concise:

Dear Flash and Spark Electricians
In reply to your quote – no thanks – too expensive.

but not designed to win you many friends or even repeat business. Try to get everything down on a single sheet of paper. Cut the waffle but don't make it too abrupt. In terms of sticking to the point remember to:

- provide a clear heading/title- this is a signal to your reader/listener;
- state the aim of your communication clearly.

(8) Does it cover what it should cover?

As we aim to be concise we have to be careful to make sure that all the key points have been covered. Isn't it frustrating to open that envelope and to find that only three of the four pages relating to the

job that you are applying for have been enclosed, or that part 1 of the quote is there but there is no sign of part 2. Even worse is the situation where the firm has at last paid up but inside the envelope, believe it or not, their cheque is for only half the amount due!

When it comes to sending quotes being comprehensive is crucial – you will stand to lose a great deal of money not to say the faith of your customer if the quote you submit leaves out great chunks of work which needs to be done!

Have you ever had a car serviced and asked the garage to carry out a check on that rattling noise coming from the front offside wheel and also to check the exhaust because you thought it sounded a little 'throaty'? You collect the car. Yes they've done the service and checked the exhaust but no they've forgotten to check the wheel. You feel annoyed. You begin to doubt the efficiency of their work. You might ask yourself, *'Well if they forgot to check that, did they actually carry out the full service?'*

So always check that what you do covers what you've been asked to do.

(9) Is it possible to get a response?

There's little point in just hammering away at communicating with people if you don't have a clue if anyone is actually receiving, listening and attending. Effective communication is about checking that the message has gone home. This means taking trouble to find out if there is a response, and if there isn't one then asking for one – in the nicest possible way. You have probably been at the receiving end of remarks such as these:

Well I thought she'd understood that we have to scrap it.

As you never said anything I thought it was OK to go ahead and knock it down.

We can help this feedback process if we encourage our readers to make use of a tear-off reply slip which with a couple of folds along the dotted lines, makes itself into a pre paid envelope. We can make sure that we've included our telephone/fax/e-mail numbers and that our address is clearly printed. We can even do some gentle reminding a couple of weeks after sending out our leaflets, etc. *Gentle* reminding is recommended. If it is too brash and harsh then people will just put the phone down. See pages 141–5 for more detail.

So many problems in communication arise because we don't bother to get feedback/response to our message. We often assume that because the message has been sent all is well. It is a very dangerous assumption. An engineer at a signalling headquarters was responsible for monitoring the way instructions were being carried out down the line. He had assumed that the new rules for signalling had been received and were being observed. This assumption was false; 35 people died when a signalling failure caused by a loose wire, which should have been checked, resulted in a train crash. We should always take care to find out that the message as sent has in fact been received and acted on. **Never assume. Check**.

(10) Is it OK for tone?

It's not so much what you say as the way that you say it. You know the phrase. It is so often that curt tone of dismissal in a letter that leaves the reader feeling cold, rejected and angry, e.g.

Your quote has not been accepted; the standard of presentation was poor.

Hey wait a moment!

You won't be needed on this job – you're surplus to our requirements

Thanks very much – how nice after all those months hard work!

Tone of voice, tone in the writing can make a very big difference to how we feel about the communication, and the communicator! Some researchers suggest that only about 20% of the message is communicated though the actual words used; the rest is the tone of the voice, the expression and gestures used to convey the message. This is what really matters. We're not suggesting that you can't be frank and to the point when dealing with others, just that unless you can put yourself in the shoes of the other person you run the risk of hurting them or causing offence. A hurt or offended customer is not going to come back. A hurt and offended customer is going to tell friends and neighbours, as in 'I want to warn you about Bloggs Builders, don't touch them with a barge pole'.

You may find the words parental, adult and childish being used to describe how people communicate. This refers to a way of looking at communication called *transactional analysis*, TA for short. This isn't the place to go into detail – at the end of this book we list some follow up material which will include TA.

Basically TA states that if you use a parental telling off, 'I'm telling you' tone of voice and point the finger this will tend to trigger off the 'child' in the person being talked to and he/she will tend to behave in a childish way – kicking the furniture, slamming the door, shouting out, swearing, or worse. Such parental tones remind them of how it was when they were spoken to by a parent or parental figure such as a teacher. What we should try to do – and it is very difficult – when spoken to by a 'parent' is to keep calm and cool and reply in a grown up manner as an 'adult'. We should, according to TA, learn to stop ourselves being 'hooked' into our child. We can do this by:

- counting up to ten silently before we say or do anything by way of reply;
- sitting on that letter we wrote in anger – having another look at it in the morning after sleeping on it;
- diffusing the situation by excusing ourselves and walking away;
- keeping our voice tone level and trying to stay grown up and 'adult'.

It is this 'adult', 'child' or 'parent' tone together with a warmth of feeling, enthusiasm, or lack of it, that is remembered when the actual information communicated has become blurred. We'll be saying much more about this later. See pages 132–3.

Here then are some of the basic ingredients of all successful communication. Whether you are studying at college, working for a company or working for yourself you will find that all your communication can be improved if you consider the ten key questions and be aware of them whenever you communicate.

Before leaving this section let's summarise these key questions in the form of a checklist. You could use this checklist as an individual, looking for instance at how you see the communication in a particular organisation that you work for or are studying in. You could also do it as a group exercise.

Activity

Take a sample of communications – letters, reports, notices, memos from your college or place of work – and assess them using this checklist. If you record a No against any of these in the checklist you might think about how they could be turned into a Yes.

Checklist

	Yes	No
Is this communication necessary?	☐	☐
Is it targeted?	☐	☐
Is it timely?	☐	☐
Is it in the right language?	☐	☐
Is it clear?	☐	☐
Is it accurate?	☐	☐
Is it short and to the point?	☐	☐
Does it cover what it should cover?	☐	☐
Can we get a response?	☐	☐
Is the tone OK? Adult not parental?	☐	☐

Methods of communication

Having established some basic principles of communication let us stop to look at all the various ways – the channels – by which we can communicate.

Face-to-face

Old fashioned perhaps but still the most effective in many cases. Talking to someone eye to eye, face-to-face will often ensure that your message is taken seriously. You can't so easily ignore such an encounter as you can a fax, a letter or an e-mail. You can't just put it into your pocket, or in a file and forget all about it.

Surveys of communication amongst staff suggest that what is really appreciated is when the manager comes down and talks face to face with staff as opposed to him or her simply sending a memo or pinning up a notice on the board. Face to face means that we invest some of our time in communicating. So it is a powerful means of getting our message across. In these days of increasing use of electronic forms of communication – fax, phone, video links, WEB – face-to-face may be seen to have great power. But there are many problems associated with it.

A person says they're too busy to see you. Just drop me a note will you, they say. They may well think that you're going to be aggressive towards them in a face-to-face encounter – bite their head off, make a fuss in front of colleagues. The other party may well avoid a face to face communication just because of these fears. It is up to us to reassure the other person and negotiate with him or her as to the best time and place for such a meeting. Saying, 'Have you got five minutes around 12 o'clock?' is a better technique than simply walking into the work space and expecting a one-to-one. Sometimes of course that can't be avoided, especially when there is a crisis situation.

Face-to-face encounters often need follow up. It's good to talk, as the saying goes, but talk is very perishable. It doesn't hang around. We need to record the basic facts of any face-to-face meeting before they fade in our memories. Get into the habit of jotting a few points down so that your memory of what was said can be matched to that of the others.

Do remember that while you're talking to someone face-to-face, he or she can't be doing anything else. You, by coming into their space, are preventing them from getting on with their work. Be respectful of this and watch the time. If you say, 'I'd like a couple of minutes' then stick to it, otherwise negotiate for a longer time. e.g. 'What I'd like to talk through will take about ten minutes. Have you got any time tomorrow?'

Non-verbal communication

One of the benefits of being face-to-face with someone is that you can pick up all the non verbal communication – the facial gestures, hand and body movements that are exchanged. This is called *non-verbal leakage* – all the gestures, facial expressions, etc. which are given off. Then there are the sounds in the voice – the tunes, the way that we express our joys and sadnesses through the way we sound (see pages 35–6).

Those who study communication reckon that 50–85% of all communication is non-verbal. If that sounds surprising then think back to

a conversation you had recently with someone you know quite well. Do you remember what was said, the actual words used or do you remember the way it was said? e.g:

• the warmth
• the humour
• the sadness
• the anger
• the disappointment, etc.

All these emotions and feelings are expressed through smiles, laughter, twinkles in the eye, heads thrown back, slumped shoulders, etc.

The non-verbal channel of communication is then a very powerful one. We cannot afford to ignore it. Some people are better than others at picking up the non-verbal cues, the leakage. How good are you? See page 35.

Meetings

These can vary between small-scale informal chats over mugs of tea to formal affairs with set agendas and rules of procedure. Most of you reading this will be involved with the more informal variety. Beware, these can waste a great deal of time. Here are a few suggestions.

• Make sure you know why you are there.
• Make sure that the aim of the meeting is made clear to all present.
• Try and make sure that a time limit is put on the meeting.
• Suggest a finishing time.
• Suggest a list of items to discuss so there is no waffling around.
• Get the top priorities in first and then, if there's time, the rest of the business can follow.

It so often happens that crucial time is spent over fairly trivial things

that could be delayed until another time, then there's precious little time left for the really important issues – like bonuses!

The point we made above about jotting down the key points at the end of a face-to-face meeting is even more important when you have a whole team meeting together. You can imagine the potential for confusion if 10 people at the meeting go away with 10 different recollections of what was said and agreed. That is why we have minutes/notes of a meeting (see pages 87–9). We'll be supplying more information on meetings on pages 80–87.

Telephone

The telephone is a very efficient means of communication and one that is getting cheaper; however, it is often not that effective. This lack

of effectiveness is frequently because the speaker may rush the phone call – his or her mind being very much on another job rather than on the actual call. It may be that the person taking your call is right in the middle of doing all kinds of other things and so your call is very intrusive. One useful tip is to ask if this is a good time to call – say you can phone back later – negotiate a time suitable to both of you. This gives the respondent some time to collect his or her thoughts and come back to you in a more positive state of mind. As with face-to-face communication and meetings it is vital that you take a note of key points from the call. Get into the habit of jotting them down.

Mobile phones are very useful. However, don't make the mistake of advertising the number of your mobile phone for others to call you on only to leave the thing permanently switched off. (That's OK provided you subscribe to a call back system which will record your messages.) Remember you'll get charged for that service.

Remember as well that a mobile phone is likely to ring at any time; you might be in the middle of a difficult job. You don't want to have to snatch up the phone; if you do that you may well often snap at the caller – that could be really good for business! If someone phones you on your mobile while you're really concentrating hard on a job you'll probably answer while you're all hot and bothered or be off-hand and abrupt. Allow yourself to cool down and collect your thoughts. You certainly don't want to be forced into agreeing something that later you'll have good cause to regret. Tell them you'll phone back. Explain that you're in the middle of doing something which is difficult, dangerous and needs your full attention. They'll understand if you explain it reasonably. Cool down. Get yourself ready to phone back when you can really concentrate 100% on the call. This might mean getting out some notes, re-reading a letter or talking to a colleague first. We'll be saying more about this on pages 141–5.

The letter

On the positive side letters often set out clearly what is to happen; they can help clarify a situation; they can soothe troubled feelings and

help people accept bad or difficult news. On the negative side they may be rapidly written and so cause offence and do nothing whatso-ever to clarify an awkward situation. They may get swallowed up with all the other incoming mail and not be attended to for days, even weeks.

It is important to follow up letters. If you haven't heard anything you might try phoning after a couple of days (assuming you use first class stamps) to ask if the letter has been received and if there is to be any follow up.

With really urgent mail use Recorded Delivery. It is more expensive but you'll know that it'll get there faster and that it will reach the actual person it is intended for since a signature has to be obtained from the recipient.

Graphical communication

A picture's worth a thousand words – so the saying goes. Well that will depend on the picture. However, we can say that in many cases a drawing, diagram, flow chart, graph, bird's eye view, sectional draw-ing, sketch, plan, layout, etc. can be very helpful when it comes to communication.

The Stationery Office have a sign catalogue which lists hundreds of signs. These can be classified under:

- *General information*

GOODS INWARD

- *Warning signs.* These give warning of potential risks.

DANGER HIGH VOLTAGE

- *Mandatory signs.* These require actions or activities that will contribute towards safety. They are printed in blue on a white background.

Automatic
Fire Door
Keep Clear

• *Safe condition signs*. These indicate escape routes in the event of fire or emergency. They are printed in white on green.

```
┌─────────────────┐
│ Fire            │
│ Assembly √      │
│ Point           │
│                 │
└─────────────────┘
```

• *Fire equipment signs*. These are used in the location of fire equipment. They are printed in white on a red background.

```
┌──────────────┐
│ Fire         │
│ Blanket      │
└──────────────┘
```

• *Prohibition signs*. These prohibit actions detrimental to safety. They are printed in red on a white background.

```
┌──────────────────┐
│ No eating        │
│ or drinking      │
│ in this area     │
└──────────────────┘
```

It is important for anyone working in the construction industry, where health and safety concerns must rate highly, to be aware of the importance of communication through signs. This is important for at least two reasons:

• understanding them yourself so that you can avoid injury to yourself and help prevent injury to others;
• understanding where they need to be displayed on site and in the site office so that visitors, customers and members of the public can see them and be warned of hazards and advised of possible escape routes.

Graphical communication is important since it can help us see things more clearly than words by themselves can do. If you have ever been in a traffic accident and tried to sketch out the where-abouts of your car and that of the one which collided with it you will appreciate the advantages of producing a diagram as a way of communicating the facts. For many people graphical communication makes much more immediate sense than reams of sentences and pages of text. We'll examine this area of graphical communication on pages 171–8.

Fax

This is increasingly used. It is very helpful when you've just had a telephone conversation with a customer if you can fax the key points of the telephone call – the contract price, date for completion, etc. This is where fax really comes into its own – none of that waiting for the letter – the confirmation comes out of your fax machine within minutes of the end of the telephone call. Remember that simply sending a fax is not a 100% guarantee that it has been received by the person it is intended for. If it is really that important then ask for a confirmation fax or phone to check it has been received.

You will find that more and more of your domestic as well as commercial customers have fax numbers. You can fax them quotes and they can fax you their response. Do be sure to leave your fax machine on. And do make sure that there is some paper in it! (Have a spare roll ready to insert. Murphy's Law suggests that the paper will run out as the most crucial contract is being faxed!)

e-mail

This is spreading very quickly; it used to be the preserve of colleges and large companies but is now being increasingly used by smaller organisations and ordinary households. Its main advantage is that if you have a computer and a modem you can send e-mail literally anywhere for little cost. It saves having to go to a fax machine and buy special fax paper.

One of the challenges of e-mail is that so much is being sent that many people hardly glance at all the masses of text on their screens – the trivial gets mixed up with the important. If something is really important then do use e-mail but also send it as a fax or make a telephone call. You really shouldn't rely on just an e-mail when it is crucial that the other person receives and understands your message.

The WEB will increasingly be a very important part of all our communications – with suppliers, customers, officials, colleges and lecturers, advertisers and colleagues in the industry. Potential customers will browse through the WEB pages under Construction Firms, Plumbers, Joiners, Roof Repairers, etc. and your WEB page hopefully is where they will stop browsing and start reading and buying. You may need to seek help in having your WEB page designed; it won't be like the Yellow Pages, you'll need something more eye catching. In the next few years the construction industry will be taking this development very seriously. Keep your eye on the WEB's rapid progress.

These then are some of ways in which we communicate. We will now examine what particular barriers there are to our communication and how best we can get over these.

We will also examine our personal styles of communication and how we can be more effective as communicators. This is what we move to now.

Further exercise

Individual
Without turning the pages back jot down the ten questions we should ask about a communication.

Group
Collect in your group examples of various channels of communication – e.g. letter, fax, leaflets, etc. from the construction industry if you have them or from your college, local firms, shops, garages, hospitals, etc. Try to assess them as

   Excellent   Good   Fair   Poor

Be prepared to defend your decision to other groups and your tutor.
Provide reasons why you would select one as excellent or poor.

2

Particular communication barriers and ways of overcoming these

We've looked at some of the key aspects of communication but as you know it is not as simple as that; communication often fails and messages don't get through. There are many barriers to communication. This chapter explores some of these. It suggests ways in which we can break through these barriers.

Activity

But first think of some barriers to communication where you work, your firm or at college. Jot down the most common ones. Think of particular examples.

In this chapter we cover ours. Compare them with yours.

The systems don't work – I can't find the address

Perhaps this is the most common reason why communication breaks down. No-one can find that piece of paper, that letter from the customer:

Harry's got it in the van and we need it here in the office

or that estimate.

Where did you last see it?
It was on this desk I swear, someone's picked it up.
Are you sure?
Dead certain.

Or that scribbled telephone number

Well it was here I know it!

Do we need to continue? We're sure this is familiar, probably too familiar! The question is how do we prevent these foul ups from occurring? Here is a checklist which you might find useful.

Activity

Just before you read ours jot down a few ideas yourself, then compare your list with ours.

- Do you keep an up-to-date list of key names and addresses?
- Is this list held on paper as back up to computer files? (see pages 100–101)
- Who is responsible to keeping this list up-to-date?
- Is there a message board at the front of the office in a prominent place where important/urgent notices can be placed?
- Does someone have responsibility for keeping this up-to-date (i.e. taking old messages down)?
- Do you have a message pad so that when a call comes in you can jot down the name of caller, time message taken, etc. or do you just use any old scrap of paper, and where do these scraps go?
- Is there a system whereby those going out on jobs have an effective way of reporting back to the office? Are lunch times covered? Are change of shifts covered? Are holidays covered?

Here then are just a few key aspects to consider. Getting your communication systems right and keeping them right are vital elements of attracting customers and keeping them.

Good communication doesn't just happen, it has to be planned for and worked at. It's similar to health and safety – it won't just come by magic.

No-one wants to take responsibility (it's not up to me mate!)

You will have had the experience of sending a letter and only having it partly answered. You know the sort of thing. You phone to ask for the refund you were promised on that cancelled holiday which hasn't arrived:

> *Hello. My name's Bill Jones. I was promised a refund on a holiday that was cancelled because of an airline strike, well that was last month and I haven't had anything yet.*
> Sorry Mr Jones but we don't deal with refunds here. You should phone this number . . .

So you phone this new number, only to be told:

> Mr Jones you'll need to fill in a form to claim your refund.
> We'll send one to you. Can I have your address?

Now as you've already filled in one form and that was a month ago you are not exactly pleased to hear this news.

> *But I have already filled in one of these. I sent it four weeks ago.*
> Well I'm sorry Mr Jones I'm not the person who normally deals with these complaints. It's Sandra and she's on holiday and won't be back until . . .

By this time you're really hopping mad. We'll leave the rest to your imagination.

What we have illustrated here is a very important barrier to communication – not wanting to take responsibility. It's not my job. Don't ask me. Find someone else. I can't be bothered!

Much of our attention in this book (Chapter 5) will be on your communication with your customer. It is absolutely essential that where *you* are communicating with a customer *you* take responsibility for it. This is true if you work for yourself or for a firm. In a firm you may well share that task with someone else but whatever happens you will need to take responsibility. Don't be tempted to pass it on to someone else. Don't provide lame excuses; e.g.

I don't know anything about this.
I don't normally do this work.
I was on holiday when this was organised.

The customer will not be very impressed with your organisation if he or she hears these excuses. Your reputation and that of the company you work for will suffer.

Most organisations these days are very keen to point out how interested they are in helping their customers, how the customer is king, the most important part of their work, the only reason they're in business, etc., etc. They send staff on Customer Care courses, have mission statements such as Our Customers are Our Priority framed on their office walls. The problem is that when it comes to handling messages and taking orders then the reality is often very different. If you work for such an organisation try not to copy others' attitudes. Set your own high standards when it comes to communication. Set them and stick to them.

Some firms have elaborate checklists for their staff with respect to communication. For example:

Telephoning

All telephone calls from outside must be answered within three rings. When answering the phone you must say:
    Your name, e.g. 'Hello. This is John speaking'
    then that of the company, XY Builders
    and, 'How can I help you?'

There is nothing wrong with this approach provided that the advice is sensible, i.e. there is not much point in rushing to answer the phone within the stated three rings if by the time that you get there you're out of puff, you're swearing under your breath (hopefully) because you've just knocked your knee against the desk in your rush!

You see our point. With all this happening you are unlikely to be in a position to be able to answer the call in a calm, polite and reassuring manner. It would be better to try and answer the phone within a reasonable time rather than the required three rings; better to sound calm and pleasant on the phone to your caller than breathless and angry; better also from a health and safety point of view since there have been too many accidents with people rushing to get off ladders to answer phones.

We are not against checklists and advice to employees on communication but it is important that any such advice is reasonable, practical and makes good sense to your customers. It may be that they don't want to be greeted by Hello my name is John. They might prefer to have the name of the company first as in Smiths Builders. How can I help you? as a reassurance that they have got to the right number.

People don't want to hear a 'robot' giving out a tuneless, expressionless, characterless response. They'd like something recognisably human and friendly. We will be dealing with telephone calls later on pages 141–5.

The barrier of communicating when we are unsure of our roles

Imagine trying to act a part in a film and desperately trying to remember your lines – even the role you're supposed to play. It is very difficult to be a confident communicator if you are unsure what to say and what to do.

It is all very well expecting people at work to answer the phone, write letters, deal with customers, and generally use their initiative as communicators, but they are much more likely to do all this with confidence if they understand their roles and responsibilities. This is why job descriptions are so important. If you have one make sure that you understand it. If you don't then you really need to be assertive (see pages 54–5) and ask questions as to your role and responsibilities, questions such as:

- is this part of my duties?
- can I clarify that you want me do this?
- do you expect me to see to this every morning?
- can I just run through with you what the system is for this?

There is no merit at all in staying confused and in a fog of doubt as to what is expected of you. If a supervisor has expectations of you as a communicator and you are unaware of this then there will be disappointments; these will do you no favours.

> Well I don't think much of our new recruit. He never answers the phone or gets those letters filed. After all it's all part of his job.

Question: has he ever been told what he's supposed to do? Has he ever asked?

Check through your job description, if you have one. Take an opportunity to talk it through with your manager, site supervisor, etc. Ask questions about their expectations of you as a communicator. Don't leave any doubt in their minds, or in yours!

The barrier of communicating in a group

Think of the difficulty you might have had in expressing your views when you've been in a group who have taken a different point of view from yours. It is often very difficult. There you are shouting for your team and those round you are shouting for theirs. If there are enough of them you'll probably keep quiet. This is what we call group pressure. It forms a very real barrier to communication. It may not matter that much at a football game (unless you're supporting the losing side!) but at a meeting or within a work team it can matter very

much. If you are doing the group exercises in this book you may have found some problems in getting others to listen to your ideas. Apart from shouting louder how have you managed this?

Let's look an example of communicating in a group, taken from a building site. Suppose you wanted to lift stone material up on an electric hoist which had a limit of 250 kg. The stones already weighed nearly this amount and your mates wanted to load on many more – to save time and finish work early. What happens? Do you have the courage to communicate your views or do you shrug your shoulders and let the majority have their way, even though you know it is not a safe practice?

What can be done to reduce this pressure of the group against the individual to ease this communication barrier?

Activity

Think for a few minutes and jot down your ideas. Think back to any situation where you personally felt that pressure. Apart from pushing, shoving, swearing or just throwing something what did you do that worked?

Here are some suggestions.

- Make sure that you express your views openly and clearly; you might be able to win others over to your opinion by doing this.
- Try and provide a reason/s why you are taking a different view from the others – find a convincing example, story, case to back up your reasons.
- If you don't succeed at first then keep trying. Very often it is only when you've repeated your point of view several times that others will begin to take notice. We'll be looking at this in terms of you being assertive on pages 54–5.

Our initial view of others

Communication between people is often spoilt by the simple fact that one of the parties (or in some cases both) do not like what they see or what they hear. Their negative feelings towards the person they are communicating with produce a barrier so that much of what is said or heard is effectively blocked. For instance:

Are you here to finish that job in the bathroom? Well, hurry up and don't leave a mess like last time. We've got guests coming today.

We can't on paper reproduce the tone of voice that the lady of the house used to the plumbers arriving at her door but you can imagine it – curt, cold, sharp and unfriendly. It would be easy to take an instant dislike to such a person. It is so very easy to jump to conclusions.

In this case – a real one – the woman's child was running a high temperature and was in bed, the guests were in fact relations coming from Australia for a wedding, and the plumber who had been called out to attend to a leak in the bathroom had left the sink in a real mess. This information does not excuse the woman's rudeness but it does help to explain her curt tones and rather abrupt manner. Wouldn't you be annoyed in that situation? Would you let it show like that?

The lesson from this is, look before you leap; think before you speak; consider the situation. Don't make assumptions from a first meeting.

Let's face it. Would you like people at an interview to jump to conclusions just because you showed yourself to be nervous, sweating a bit and rather tongue tied? You wouldn't like the interviewer to think: 'Oh no, there's no way we could employ this one'. You'd hope that he or she would give you a fair chance, put aside feelings and prejudices and wait until you had calmed down, stopped sweating and breathing hard before even beginning to make a judgement about you and your suitability for the job.

We do all have our individual prejudices. Some people can't stand men wearing ear rings or tattoos, others dislike girls with bright lipstick and heavy make up, other prejudices run deeper and can have very serious consequences, e.g. race and gender discrimination which are both against the law. Here are some examples of discrimination which are very widespread.

- I'm not turning this company in to an old folks home (dismissing the application of a 55 year old experienced joiner simply from one look at his date of birth);
- Well it's not woman's work (refusing to interview a woman who applies for a job on the site);
- Bloody ear rings, not likely (refusing to employ a worker who has one ring in his ear).

It is not necessarily wrong to have our opinions about people; it would be difficult to live in this world if we didn't. The point is we shouldn't let these feelings get in the way of our communication and judgement in the workplace. How many times does a negative first impression lead to a much more charitable view after acquaintance:

Well I didn't much like the look of him at first but he's turned into a great mate to have on the job.

Activity

Consider the way in which you come over to others. Is there anything about the way you speak or in the way you appear to others which could trigger off hostile/ negative feelings and thoughts from others?

It is worth thinking about this question since some serious reflection could be very important when we come to consider your communication in the interview (see pages 51–4). Do you for instance:

- avoid looking at the person you are speaking to?
- have a curt and rather harsh tone of voice?
- speak softly and quickly so that others find it difficult to hear you?
- use certain mannerisms which distract your listeners?

You can imagine that any one of these ways of communicating (or not communicating) could be off-putting to another person. Yet you might be behaving this way because you were tired, had a hangover – that's why you're not looking at the person – and you're muttering under your breath because you've got such a sore throat from shouting at yesterday's match. If we hadn't known about these factors we might have thought that you were rude and not the kind of person we wanted to talk to or to do business with.

Putting aside first impressions

If we are going to be effective communicators then we need to be able to put aside our prejudices and feelings so that we can actually listen and talk to other people in an open and not a prejudiced way. This will be very important if you recruit people for your work. We all have negative feelings and prejudices; the important thing is

- to acknowledge them, i.e. to admit that we have them,
- to avoid letting them interfere with our judgements.

Now it is true what many people say that first impressions count. That is why we need to be particularly careful as to how we come across to others. We need to examine our own style of communication; the messages we give off. We are talking about the work place here, we are talking about colleagues, customers, suppliers, tax inspectors, health and safety officers, etc; about people who we need to be successfully communicating with and making a positive impact on. This leads us into consideration of our own communication.

Our personal style of communication

You've probably heard the saying; *it's not what you say it's the way that you say it – that's what gets results.* Well that really is true. As we noted on pages 18–19 somewhere between 55% and 85% of all our communication is non-verbal, that is to say the tone and the tune of the voice, the gestures, especially the facial ones, the *how* as opposed to the *what*.

If you think this is strange then think of a telephone call you've had with a friend; can you really remember the details of the actual message, or was it the friendly sound of the voice, the chuckles, the laughs, the emotion and expression in the voice? You'll probably admit that the how of communication as opposed to the what of communication is very important. We should therefore spend some time in examining our own style – that is how we come across to others.

Activity

Here is a short checklist for you to try out on yourself. It would be a good idea to get a friend to assist you. Tick the word in the list which best describes you. Ask a friend to put some ticks in as well. Compare any differences in your judgements!

	My opinion	My friend's
Appearance		
sad	☐	☐
friendly	☐	☐
happy	☐	☐
Voice (quality)		
soft	☐	☐
loud	☐	☐
harsh	☐	☐
stern	☐	☐
warm	☐	☐
Voice (audibility)		
easy to hear	☐	☐
muffled	☐	☐
ends of words fade	☐	☐
Eye contact, looks		
directly at others	☐	☐
over their heads/the floor	☐	☐
Posture		
stands tall	☐	☐
upright	☐	☐
slouched	☐	☐
stooped	☐	☐
Sitting position		
slumped	☐	☐
puts feet on table	☐	☐
swivels around	☐	☐
Annoying habits (e.g.)		
scratches left ear	☐	☐
rattles coins in pockets	☐	☐

This doesn't pretend to be a full list, just a few key items. But you might like to think about how you would rate yourself on these and how a friend might. Think as you complete this list how you might come across in an interview.

The point we're making here is that if we don't know much about how we come across to others then we can have little chance of putting things right. We might only be told when it is already too late, i.e. the customer who doesn't come back, the interview that leads to nothing, the friends who never phone back, the pub that empties when we arrive!

It's time to move on now to consider in more detail aspects of our personal communication.

Group exercise

Think in your group of situations where you formed an impression of a person at first sight and then changed this opinion when you got to know him or her better. Think about the situation (i.e. where you met), your initial reaction and your later views.

Discuss your findings with other groups and your tutor.
Try and analyse why your initial reaction was right or wrong?
What does this tell you about our prejudices?

3

Your personal communication

In this chapter we ask you to have a look at yourself as a communicator. This is not easy. We have all got rather used to ourselves – the ways we speak, listen, gesture, write, use the telephone, etc. In order to improve and develop ourselves as communicators we do need to do a little bit of self examination. We start first with the interview. For many of you that will be interviews for a job; for others it will be interviews to borrow money from a bank or building society; for others it will be getting start-up grants for a business. In the following section we concentrate on getting that job but remember many of the points dealt with will apply to the other kinds of interview.

Getting that job

Your selling points

Consider selling a motor bike or car. You know roughly what the market price is, well within a few pounds; at least you know what a ridiculous price would be – too cheap or too dear. You know the good

bits of the machine; you also probably have a fair idea of its weak areas. You do your best to tidy these up. A lick of paint here and there. Some thorough cleaning away of grease marks would help but more serious defects will need more substantial treatment.

When you apply for a job you're entering a market. You have your value; so do others; and so does the employer. He or she will look at your application and hopefully invite you in for interview. If you measure up you'll most likely get the job. This chapter is all about helping you get to that position.

You have to get yourself ready for the market. You need to think carefully about your positive points – how you can enhance these, make sure that they're not overlooked. Then there are your less positive points – it might not be enough to give these a quick going over. These might require something more substantial in terms of 'rust removal'. To start with you need to take a good long and frank look at yourself from an employer's point of view – that's the perspective that counts.

Here's a balance sheet for you to think about. You will find this helpful. It is better to do this before we go on to explore application forms and interviews. This is the starting point, a self inventory, a SWOT exercise – your Strengths, Weaknesses, Opportunities and the various Threats to these (see pages 115–17 for more details). This is all about asking the question, do you know your real worth? As with our bike or car there's no point going to the market unless we have some idea as to our value and the 'price' that someone might pay. We have to be realistic but at the same time optimistic. So let's first of all have a look to see what you have in the 'bank' so to speak.

Activity

Before you read further think what your strengths, your positives, are as far as an employer is concerned. Jot down a list.

Here are a few to think about.

- Experience. A crucial issue. What kind do you have? What might an employer consider interesting/relevant?
- Qualifications. Important. What do you have? How recent are they? What level are they? Do they come with any credits? Are they widely recognised?
- Skills. Some of these will go under qualifications, others will relate to skills that you have gained not from college or study but from experience, your own initiatives, your sports, interests, etc.

- Personal qualities. These are considered of great importance by employers: honesty; reliability; a good even disposition – you don't go mad with temper and fly off the handle for very little reason; a sense of humour (invaluable when you're working with a team); punctuality (again vital as a team member); good attendance, i.e. not prone to many days off, etc.

Well that's just a starting point; we'll be going into more detail when we come to interviews, etc. But can you see just how important it is for you to do this kind of exercise? An employer won't do it for you. A friend can be helpful and so can your tutor at college. If you don't get any feedback on these important aspects of your work and your qualities then ask.

You will need to be able to answer that very important question from the employer: what can *YOU* do for *US*?

Telephoning for a job

This is increasingly the way that jobs are gained. You see an advert in the local paper and then decide it is for you and hurry along to the phone box. Wait! Don't hurry, that's our advice. If you do then it is likely you'll get yourself into a muddle. Now that won't matter too much if the telephone call is simply to secure you an appointment to see someone; however, it is quite likely that the person on the other end of the phone – listening to you hurry and muddle your way through the call – will be the potential employer. He/she won't gain a very favourable impression of you if you gabble your way through the call in a unstructured and badly thought out manner. If you sound breathless and hurried this will not gain you much credit.

Activity

If you have made a phone call for a job think back to how it went for you.
After your experiences what advice would you give to a friend who was calling?

Here are a few tips from our experience and from talking to employers.

- Take it slowly. Plan for that call. Read the advert very carefully. Look out for key words such as experience, qualifications, availability, location of work (how will you get there? public transport?), and obviously starting wages and conditions. Remember that what may be stated in terms of wages may be negotiable if you can prove to be the person they really want. Be optimistic but be realistic!

- If possible find a phone where you won't be interrupted. It is very difficult to sound cool, calm and collected on the phone if you have people hammering on the outside or stomping around impatiently fuming while you conduct your delicate negotiation. The other problem with a public phone is money – it's crucial that you have enough so that you avoid having to pump in coins just as the conversation is beginning to get somewhere. Avoid using a mobile as connections can so often be poor and intermittent.
- Try and find a place where you can sit down at a desk and lay out the advert and some notes to assist you in making the call.
- Make sure you have really read the advert carefully – sorry but you'd be surprised how many don't.
- When you phone try and establish quickly who you are and what you are phoning for.

Good morning. My name's June Smith. I'm phoning in response to your advert in the paper.

There's no point in saying much more than this, certainly little point in starting your great sales pitch until you have established whether this person who has answered the phone is the one you really want to speak to. You know the sort of thing – you've been talking for a couple of minutes, giving of your best when the voice at the other end says:

Sorry I think I better pass you on to Mr Jones; he's the one dealing with this.

This is so annoying. Declare who you are and what you want and then let them find the right person. Remember you're phoning in unannounced (although you would hope that following the placing of the advert the organisation would be expecting some calls!). Be prepared to phone back. A little bit of negotiation here could create a good impression:

This might not be a convenient time to call. I could phone back later.

With some phone calls the actual interview is done there and then. So be prepared to have those essential pieces of information from the previous page at your fingertips. This is certainly not the time to waffle. You need to sound confident, unlike this example.

Hello, yes, I'm phoning about the advert (which one?).
I've got my certificate in plumbing (which one?)
and had some experience (how many months, years?)
with a local firm (who?)
and I'm very keen to start (aren't they all)
as soon as possible (does that mean you're not in employment at the moment – funny?)

You can see the problems this caller is running into. His or her credibility is running rapidly away. Compare this.

Hello. My name is June Wright. I'm calling about the advert for a joiner in today's Post and Telegraph. I've got my City & Guilds . . . and had 18 months experience working for Hall & Thomson. I would be able to start at the beginning of December and could come for interview . . .
Yes I can be contacted during the day at . . . and after 5 at . . .

This is a more credible performance. Remember:

- do try and speak as clearly as possible;
- give your name clearly (remember the caller may well be listening to you in a noisy room with other people talking);
- spell out your name if necessary;
- say your telephone number and address very clearly;
- repeat the number if possible.

It is crucially important that you leave a record of your name and a contact telephone/address. If you have a fax or can borrow the use of one then immediately after the call, fax in your name and key details, especially your phone/fax number and address. The same advice would apply if you have e-mail. We realise that for many of you this is very obvious advice. But remember those binoculars on the Titanic – the ones they couldn't find. It is often the obvious that gets forgotten.

The aim of your telephone call is to get sent an application form or to be given a time for interview/practical test or even to be offered that job! Yes let's be positive. We now turn our attention to that application form.

Application forms

The first thing to do when you get an application form which may look very daunting and off-putting is to go to your nearest newsagent or stationery shop and get a photocopy. Place the original form in its envelop flat in a drawer and start working in pencil on the copy. Take your time. These forms have usually been very cleverly designed to catch as much information as possible. They're not to be rushed. Here is some advice.

- Don't leave any blanks – a blank might mean (a) that you've forgotten to fill this part in, or (b) that you have something to hide. If necessary write Not Applicable in the space.
- Keep to the boxes provided. That's why there are boxes! Avoid using arrows to indicate that this bit links with that. Leave the tech drawing till later!
- Check that you are in fact answering their question and not the one you would like them to have asked (very tempting)!
- Take the greatest care with spelling, punctuation and grammar (see kit bag at the end of this book, pages 152–92). Don't trust your own checking skills; show the photocopied version to someone else and get a fresh pair of eyes on it. Don't make mistakes copying from this version to the final one. Triple check that you've spelt the name of the company correctly, e.g. Morrisons The Joiners, not Morisons Joiners.
- Check your facts. Don't be tempted to exaggerate your skills, qualifications, experience or date of birth. Most employers will make checks on your application. The truth will come out and if your truth's wrong – you'll be out!
- If you need to provide a photograph to accompany the application, then don't be tempted to enclose the one you had taken with some boozy pals with you in that baseball hat! Get some more done at the post office or train station where they have those machines. Look the camera in the eye and try not to blink each time the camera flashes.
- Where the form asks for further information about yourself don't leave that blank. This is your opportunity to sell yourself; take the chance with the photocopy to practise writing a paragraph that will provide the reader with additional information about yourself which will tempt him or her to call you for interview or even give you the job. Don't waste it with a couple of lines like this:

    I am a good worker and look forward to working for your company.

This is worse than useless. Think of those personal qualities which we considered at the start of this chapter. This is what you should be putting down:

> Apart from my work at college I have also been involved with ... club. I am a keen ... player and turn out most Saturdays. I would describe myself as .... During my college course I took part in ... and gained the ....

- Make sure that you sign and date the form. Take a copy and keep it. This will be useful next time you have to complete a form. It will save you time as many of the details will stay the same.
- When you have done all your practising then transfer the material on to the original application form.
- Use black or dark blue pen, avoid light colours and shades, these don't photocopy well. Never use pencil.
- Place the completed, signed and dated (and double checked) application form in a large envelope.
- Try and find some cardboard backing so that it doesn't get crumpled.
- Write a covering letter.

Activity

Have a close look at this extract from a completed application form.
What advice would you give to the applicant on how it has been filled in?

---

1. Full name (BLOCK CAPITALS)   Shirley J Smith

2. Address   14 High St
                                  Elborough

3. Telephone   0112-212-234

4. Present Employment   Elborough College of
                                  Building

5. Skills/Qualifications (with dates)   NVQ level 1

6a. Previous employment   Hard Up Builders –
    (most recent first)   Summer 1997

```
                              Up and Away Builders -
                              Summer 1998

6b. Main duties/responsibilities   Helping generally
_____
7. Interests                  Fishing
_____
8. Referees                   Mr David Thomas
                              Hard Up Builders
                              16 Holly St
                              Elborough
_____
9. Signature
```

Here are some of the mistakes. How many of these did you spot?

Section 1. If the form says block capitals then print your name in BLOCK CAPITALS.

Section 2. Always put your postcode under the address.

Section 4. Shirley is at college. She should state this and what course/s she is doing.

Section 6a. It does ask that you put your most recent job first. If the form requests you to fill it in this way then do so. These two jobs are temporary summer employment. Shirley should state this.

Section 6b. Helping generally is rather vague, to say the least. Try and be specific. Remember this form is your shop window. If you don't put the goods in the window a potential employer won't be able to see them!

Section 7. Try and say more than just *fishing, football, TV, pubs*. Remember that window. Add some detail which provides a little more about you and what you like doing or are good at doing. River Fishing? Member of Elborough Angling Club? If you do belong to a club or organisation put it down; it tends to demonstrate that you are a fairly sociable person who gets on with others. Your membership can start a conversation going which can be helpful in encouraging you to talk and allowing those listening to hear more of what makes you 'tick' as a person.

Section 8. A referee is someone who can provide additional information about you. If you are at college then one of your lecturers will normally act as referee. He or she should be someone who has taught you or supervised you and the longer the better. Normally it is wise to put the names of two referees. The second one could be a member of the staff where you have worked before college, a school teacher, or someone who has known you as a family friend. Shirley should have provided two referees. She should also have supplied a fuller address and telephone number. She should also have indicated what position Mr Thomas is in this company. Is he site foreman, general manager, family owner, training manager, etc? Do remember to contact referees before putting their names on your application form. They may well get a telephone call or letter requesting particular information such as your reliability, honesty, health and attendance record, etc. Tell your referees what kind of job you are going for. This will help them when making up your reference.

A final word of advice before you send off that application form. Show it to someone else and get him or her to have a very close look at it. You see how Shirley in this form (**Section 9**) has missed out on putting in her signature. It is a small mistake but errors like this will reduce your chances of getting called for interview.

Activity

Here's another version. Do you think it's OK now? Read it through carefully as though you were the prospective employer.

---

1. Full name (BLOCK CAPITALS)	SHIRLEY J SMITH
2. Address	14 High St Elborough EH6 7PQ
3. Telephone	0112-212-234
4. Present Employment	Student at Elborough College of Building Attending year 3 NVQ Building Course
5. Skills/Qualifications (with dates)	NVQ level 1 Taking level 2 June this year

6a. Previous employment (most recent first)	Up and Away Builders – Summer 1998 Hard Up Builders – Summer 1997 Both these were temporary holiday jobs
6b. Main duties/responsibilities	Helping generally. Sorting out tools, sending messages, checking site notices, office duties, photocopying, filing, etc.
7. Interests	Fishing. Member of Elborough Angling Club.
8. Referees	Mr David Thomas Manager Hard Up Builders 16 Holly St Elborough Mrs Jane Taylor Senior Lecturer Elborough College of Building Hartford Rd Elborough EH12 5TY
9. Signature	*Shirley Smith*

This certainly looks better. The more glaring errors have been eliminated. However:

- Shirley might still have added a little more to the information on skills and interests – it's rather on the bare side;
- she might have added something more to 6b – 'helping generally' is a bit on the vague side; she might have added a bit more detail.

The covering letter

Although this only needs to be a short letter it is a very important one. It provides evidence to the employer that you can write; it should stimulate his or her interest in reading your application; and it should provide some evidence of the care and attention that you are prepared to take. This all assumes that you write a good letter which enhances and does not detract from your application!

We will be looking at letter writing in some detail on pages 152–5 but here are a few tips.

- Make sure you write the letter on good quality paper – letter writing paper and not any old sheet of paper.
- Take care to address the letter to a person – you should find information on this within the advert. If there is none then you could try Personnel assuming it is a big enough firm to warrant one or simply address it: To Whom It May Concern.
- Don't just copy the material you've already included in the application form; highlight what you consider to the be the most important. This is where you have to try and think what, if you were in their shoes, would prove attractive – your qualifications, experience, etc.

Activity

Imagine you are an employer. Have a close look at this letter of application. The numbers in the text refer to the next page.
Would you, on this evidence, want to see the candidate?

Original letter

	Flat 3
The Manager	34 Castle St
Appletree Builders	Linchfield
Marsh Lane	L7 1UP
Linchfield	
L5 7PS	*(1)*

Dear Manager

   In response to your advert I enclose an aplication form. *(2)* I gained my NVQ *(3)* in 1997 and for the last 15 months have worked with James Hewitt Builders. I have gained experience in general building. *(4)*

   I am interested in working for a company such as yours. *(5)*

Yours sincerely (6)

Mike Smith (7)

Well what impression would you have of the applicant from this letter? Does it say enough about him or her to sharpen your appetite or would you put it down in the don't call pile. What could be improved?

(1) Date – always put a date on a letter.
(2) Double, treble check your spelling – e.g. application.
(3) Don't be vague – supply details. Remember many employers have vague ideas when it comes to qualifications – spell them out, i.e. what level of NVQ? Remember if applying to a Scottish based firm that they are more familiar with SNVQs.
(4) What experience? One year or six weeks? This information is so vague as to be almost worthless.
(5) Too obvious! Try and write something more original.
(6) If you start with Dear Sir, Dear Manager then stick to the formal ending Yours faithfully. If you use a name, Dear Mr Smith, Dear Harry Smith, then use Yours sincerely.
(7) Please print your name underneath your signature. How is the reader going to be able to decipher this autograph?

Apart from these obvious mistakes, there is a need in the first line to say where you saw the advert – the newspaper, job centre, etc. Also notice the repetition of *gained* in lines 1 and 3. Try and avoid using the same word; find a substitute – achieved, obtained, was awarded, etc.

Activity

Let's assume that the letter of application avoids these errors. Is it now good enough to send? Give it a careful read through.

Second attempt

```
                                    Flat 3
The Manager                         34 Castle St
Appletree Builders                  Linchfield
Marsh Lane                          L7 1UP
Linchfield
L5 7PS
                                    Sept 30th 1999
Dear Manager

    In response to your advert in the Herald on
Sept 21st, I enclose an application form.
    I gained my NVQ level 2 in 1998 and for the
last 15 months have worked with James Hewitt
Builders in Elborough. I have gained experience
in general building. This included scaffolding,
brick laying, slab laying and roofing. I also
assisted with two central heating installations
on new housing developments in Elborough.
    I am available for interview and can be
contacted on 0113-456-678.

                    Yours faithfully

                    Mike Smith
                    Mike Smith
```

It is certainly better, but how could it be improved?

- He could have written a heading at the top of the letter. This would have immediately alerted his reader as to the topic. e.g.

Application for Joiner's Post

- He needed to add *where* after Elborough as in *where I have gained*;
- He could have avoided using *gained* twice by substituting *awarded* at the start of the paragraph as in *I was awarded my NVQ Level 2*.
- He gives a number where he can be contacted. Is that telephone or fax or both ? A small point but worth clarifying.

Now for the interview.

The interview

To put it at its simplest: an interview is an opportunity for you to find out more about the job and if you'd really want it and for the employer to size you up, to ask questions resulting from his/her reading of your application. You may think it a bit strange that we've started with the idea that the interview's for you to find out if you want the job, but if you think about it for a moment it does make sense.

There's absolutely no point in your going to the trouble of filling in the application form and then going for interview if during this process you discover that this job is not for you. No matter how closely you read the advert and any other information that is sent to you it is very often only when you talk face-to-face with the employer-to-be that the true nature of the job becomes clear.

As you talk and listen in the interview you must be prepared to say *no*. It really won't be in your long-term interest to say *yes* to a job that you can't actually do, or that you will be bored by after a few weeks – unless that is you are really desperate to get this work, or you feel that, though it is not what you're really after, it will provide a stepping stone to something better.

Activity

If you've ever been interviewed for a job then think back to what it was like. What advice would you give to a friend who was due to go for one in the near future? Jot down your top hints.

Here are some.

- Go early. Take good care to find out where the place for interview is. Don't assume that it will be held at the address indicated at the top of the letter. Phone to check. It is very difficult to feel relaxed and give of your best if you're running late and rushing.
- By going early you will be able to look around and get the feel of things. Take the time to see what's happening, what kind of organisation this is, i.e. the range of work they carry out. If there's a leaflet lying around do read it. You might be in a better position to sound alert and on the ball at the interview.
- Try and make it as much of a conversation as possible. Don't get into the yes/no answering game. If all you give back is a mixture of these two words then you can effectively say good-bye to any chances you might have of getting the job. Provide full answers; provide examples of what you have been doing. i.e. not

I've been working with Smith the builders for 2 years.

but

I've had two years with Smith the builders; in that time
I've had experience with ... and I've also done some ...

**Remember if you don't put your 'goods' in the shop window then
others won't be able to see.**

- Don't wait for the employer to squeeze the information out of you;
  provide some yourself. Try and build up a picture not only of what
  kind of experience/skills you have but of the kind of person you
  are. Avoid this answer to the question:

What do you like doing in your spare time?
Nothing much, watching telly, going out with mates.

Try if possible to provide a slightly fuller picture

Well sports, watching these on TV. Playing the odd game with
mates.
I've been in a few pub quizzes – not very successful but it's fun.

Remember just how much communication takes the form of non-
verbal expression (pages 18–19). So do:

- *Sound* interested. Put a bit of zest into your answers – enthusiasm
  will often carry the day. You may not be top of the pile when it
  comes to experience but you may be able to hoist yourself up there
  if you come across as interested, willing, with oomph!
- *Speak up.* You may be interviewed in the yard, on a site, in the
  middle of a busy office. You might have to contend with all kinds of
  noise. If the interviewer can't hear you or has to ask you repeatedly
  to speak up then your credit rating will drop ... rapidly.
- *Look at the interviewer* not at the floor. We don't mean fix the guy
  with a rigid stare but please don't spend all that valuable time when
  you should be making a positive impression looking at the carpet.
- *Avoid fidgeting.* Keep reasonably still. We don't mean that you have
  to sit or stand fixed without motion – that would only increase your
  nerves and the employer's, but go easy on the head/armpit
  scratching, hopping from one leg to the other, playing with the pen
  (especially the ones with click tops!) or fidgeting with collar,
  trousers, etc.

- *Get comfortable*. Pull the chair back so that your legs don't knock against the table. If the sun (or that lamp) is shining into your eyes then move; don't get stuck in one place where you have to screw up your eyes and squint at the interviewer.

As far as what to wear is concerned the golden rule is: wear something that makes you feel confident. There's no point in dressing up if what you dress up in only makes you feel uncomfortable and awkward; it must look clean and smart.

If despite all this advice you don't get the job then it could be for all sorts of reasons.

- Perhaps they were looking for someone with more skill or more experience.
- Perhaps they thought that they could find someone better.
- Perhaps you just weren't suited for the job in some way or other.
- Perhaps they already had someone for the job in the firm but just thought they'd see who else might be around.

There are all kinds of 'perhapses'. In view of this we would advise you straight after the interview to find somewhere quiet and have a little think to yourself while the memory of that interview is still fresh in your mind. Ask yourself:

- Where could I have done better? What different answers to particular questions?

- Were there missed opportunities to provide examples? Was there a failure to speak up?
- What were my strengths? How can I build on these?

Some companies will offer you feedback on your application if you weren't successful. Take advantage of this; you may learn something about the way you come across at interview which will be very helpful later on.

Assertiveness

It is very appropriate that right after outlining some key points on interviews we should examine assertiveness. You may have heard people say to you, if you have kept quiet when others have spoken up, that you should be more assertive! A good number of people use this word but not all of them know what it means.

Assertive behaviour very briefly means respecting your rights and those of others. At one extreme is passiveness which means someone else has all the rights and you have none and at the other extreme, aggressiveness means that you have taken over all the rights and the other person has none.

Passiveness	Assertiveness	Aggressiveness
I don't have any rights	I have rights; so does the other	I have all the rights
I'll keep quiet	I should say something	I'll say what I want
It's not my place to say anything	It's important my voice is heard	You don't count
No-one's interested in what I think	We should all have a say	Shut up. I'll tell you
There's no point. It'll not change	There's a chance for change	I'll change it
I'm not worth listening to	I have a view so I'll express it	I'm right – you're wrong

You can probably recognise yourself (and your friends) in this table.

We can see this in the form of a see-saw with assertive behaviour in the middle and passive and aggressive behaviours on the edges. We should try and keep some kind of balance in our actions. If we start shouting and losing our temper then we move along the see-saw towards the aggressive end. If we keep quiet, don't express our feelings, shrug our shoulders and say *Well it doesn't really matter* then we are moving down towards the passive end as in the diagram below. Try and keep in the middle – be in balance. Express your wants and views in a confident and clear manner. Remember as we said on page 32 you may have to repeat your request and repeat it several times before others take you seriously.

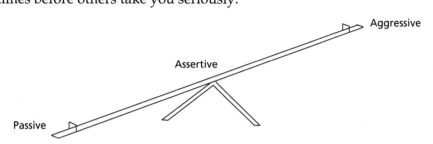

The point we wish to make is that being assertive and learning to stay assertive is a very important aspect of communication. There's no point in filling up that application form ever so neatly and looking ever so smart at the interview if when the employer says:

That's not much of a qualification is it?

you simply agree and nod your head (passively) or glare at him and shout:

What the hell d'you mean by that? (aggressive)

What you need here is some assertive behaviour:

Well it was taken part time, I was working during the week and it is only a first step. I fully intend to go on to take the next . . .

You see the difference. You have rights in this situation – i.e. not to be spoken to like that – but you don't have the right to shout back. This will only lead to a confrontation and an abrupt end to the interview, whereas in the case of the assertive behaviour you can gain their respect by standing your ground. It is a win–win situation. The interviewer will probably think more highly of you because you did respond in a firm way; he/she might have been trying to wind you up and be provocative.

You probably have been assertive in many situations but didn't think of it as that.

Activity

Think of those times in public when you expressed your feelings and views not in a passive or aggressive way but in an assertive one.

You might for example have been going out for a meal with a special friend. The food arrives, you taste it but it's pretty cold. Do you just shrug it off? Your friend's helping is fine, you don't want to cause a fuss, so why bother? This is a good example of passive behaviour. Or do you summon the waiter and say in a loud voice

This food's cold. I ordered a hot dish. What the hell's going on?

Very aggressive! This will convince your friend that you're certainly not a wimp! But it may also cause some embarrassment. You might have acted assertively by calling over the waiter and saying in a quiet but firm tone

This helping's cold. Could you warm it up? Thank you.

The result's the same as being aggressive but there's no embarrassment. Your rights have been respected. You also respect the rights of the waiter (i.e. not to be shouted at – it might not be his fault).

We'll be looking at other areas of work where it will be important for you to act assertively – as when dealing with suppliers, colleagues and customers.

Handling conflict

As part of personal communication we should spend some time examining the issue of conflict. We all experience conflict in our everyday lives. Conflict is inevitable. You could argue that conflict is much easier to handle when it is in the open rather than when it is buried. It's rather like a forest fire – easier to put out when it's on the surface amongst the branches but very difficult when it goes deep underground where it smoulders and may start again at any time.

In your work you will come across conflict with managers, supervisors, customers and those in agencies and authorities. The key question is *What can we do about it?* How can we best manage it through our personal communication skills?

We've already seen the importance of tone in our communications (pages 15–16) Transactional analysis (TA) suggests how we should avoid sounding too parental in our transactions because this tends to trigger off a childish reaction. So one way of managing conflict is to maintain an adult tone in our communication. If someone is angry with us at work and uses very heavy parental language:

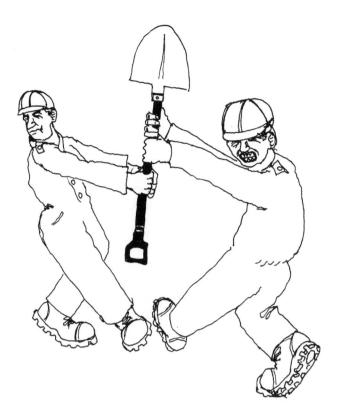

Why the hell haven't you done it?
You're just stupid.
Get off your butt and damn well see to it.

then it is fatal to use parental language back:

What the hell do you mean?
Get stuffed.

This will inevitably trigger off a war of words and both sides will end up hurt. Remember a short-term gain – you've told the other person off good and proper – will only be short term. At work we have to try to sustain long-term relationships where, though we might not like the other person very much, we have to work with him or her. A complete breakdown in relations – the inevitable consequence of handling conflict in this way – will make it very difficult to sustain long-term working. This is why it is so important to handle conflict by staying in the adult tone and by being assertive. If we are passive then this will only encourage the other person to become more aggressive as we've seen. In the face of

Why the hell haven't you done it yet?

it's best to reply

It's on my list. I shall get it done.

or

The supervisor wanted me to see to those roof tiles first. It will be done.

Don't get into the passive:

I'm sorry I haven't had the time.

Be assertive and provide a reason. Reassure the other person that the job will be completed. Remember, don't be tempted to over promise and then have to give more excuses. Don't say

It'll be done by 1 pm.

only to realise almost as you've said it that there's no way you can get it completed before the late afternoon.

Basically there are four options for us in the way we manage conflict.

Avoid

Use this for those situations where it is not important that you win or lose. If someone's having a political or religious argument, and these people are the ones you have to work with, sell to, or manage, then avoidance may well be the best strategy. You decide that it is just not worth while getting into this conflict.

Compromise

If it is possible to find a mutually satisfactory solution to the conflict then go for it. You may have to give some and they give some. We are into negotiation here (see pages 120–23). Rather than you getting all you want or them ending up with all they want it may be very possible with some good will to arrive at a compromise where both end up winners.

Look, I must have this done this afternoon. OK.
I can do it but you also asked me to check on the delivery schedules.

Well you could do that later?
No, that would be very difficult. Those schedules need to be checked today; they'll take a couple of hours.
OK. Could you get it done before 11 tomorrow?
I'll have to do some phoning to rearrange things, but yes, OK.
Thanks.

In this case both sides have given something up: you have had to spend time rescheduling what you had planned to do – that means spending extra time and effort; the other party has had to compromise in that the work won't be done until later. A great deal of conflict can be avoided in this way if both sides are willing to compromise.

Collaborate

This means working with the other party in sorting out some solution to the source of the conflict.

But I've told you, it's really important that I get to the college.
You've already been allowed time off to attend your courses.
Yes, but this is a special session. I did explain it.
No, sorry.
Here's a copy of my work schedule for that week. Could we not work something out?
Well I'm not keen. I'm really not. Here let me see the plan.
I could start earlier these two days – say 7.30 am.
Yes and work later this one till 6.30 pm.
Would that be OK then?
Suppose so. But don't ask me again. You've been given time for college and that must be it. We're just too busy for this to happen.
Well thanks for agreeing to that.

Collaboration is very similar to compromise but it involves *both parties* working together to create a solution. It's different from compromise because it is not necessary that both sides have to lose anything. Being creative in problem solving is one of the most productive ways of managing conflict. If you get stuck see if there's some way out of the difficulty.

Think sideways, think differently, be creative and see if there's a way out. Do some brainstorming; this means getting members of the group to give as many ideas as possible. The ideas are collected on, say, a sheet of paper. The rule of brainstorming is that all ideas are written down – there's no stopping to consider if they're sensible or not. Only when all the ideas have been collected is there any consideration of their practicality.

Bottom line

In some situations you really feel for all sorts of reasons that you must be assertive and stick to your position. In other words there can be no compromise and no chance for collaboration.

> *Give us a hand to move this will you. I'm running late.*
> That's going to require lifting gear.
> *Come on give us a hand. The gear's over at the other site.*
> No. It's too big.
> *Come on weedy. Let's see you lift it.*
> No, sorry. It's dangerous. The gear'll be over here tomorrow.
> *But tomorrow's no good. I want it moved now sonny.*
> No, I'll give you a hand first thing tomorrow with the gear.
> *Well if that's your attitude, don't ever expect any favours from me MATE.*
> Look I'd like to help but you can see yourself if that slipped out of our grasp it could do us some nasty damage. It might crash through and injure others below. As I said I'll come across first thing and get it done.

Notice that in this conflict situation the person refusing to take part in the lifting sticks to his position, there's no compromise. If the danger was less and the problem easier to solve then some collaboration might have been in order. e.g:

> Why don't we get hold of the lower end, get some of these planks to support it while we move it sideways? That way it'll be safer to move.

But because of the risks to their safety and to those of the others working below no compromise or collaboration was in fact possible. In these situations it is important to be assertive and hold your bottom line. If you don't you'll end up by regretting your actions and feeling you've let yourself (and others) down.

There is more on the bottom line approach to conflict in the section on negotiation (pages 120–23).

Group exercise

In your groups consider situations:

(1) where you showed assertiveness or managed conflict successfully;
(2) where you now wish, looking back, that you had shown more
assertiveness or had handled a conflict with more success.

For part (2) you might want to discuss these questions.

- What was the situation?
- How did you handle it?
- What should you have done?

Compare your results with other groups and your tutor.

4

Communication on site

Having looked at our personal style of communication in terms of interviews and being assertive we now turn to the communication we will need when we have got the job and we are having to communicate on site. This will involve suppliers, colleagues, and in the next chapter, customers.

We saw in Chapter 2 that too often failures in systems are responsible for communication lapses. We also noted that there is often an element of:

It's not my responsibility – let someone else have a go.

Both these present serious troubles as far as good, clear, reliable communication is concerned. To help combat these, procedures are put together. These aim to set out as clearly, concisely and directly as possible:

- what should be done?
- in what eventuality?
- how?
- by whom?

They are put in writing because it is well known how slippery verbal procedures are – like bars of soap in the shower. There's just too much risk of things being misinterpreted, of meaning slipping away.

Such written procedures are important in that they are not person specific, i.e. if Harry Jones is the only one who understands how the new cement mixer operates and he leaves to go on another job we need to be able to get the thing going here and now. If this is not done you can imagine the situation.

- Where are the procedures?
- Who can understand them?
- Which scrap of paper did he write them down on?

This is why we want well written, clear procedures. These should be kept in a place where everyone who needs them can get hold of them. We don't want situations where the procedures are in Harry's pocket and he's left his trousers for cleaning!

Written procedures

Let's start with something simple.

Activity

How would you write down a set of procedures for striking a match? The procedure has to start from the time you open the box to the moment you strike the match and it flares into a flame.

Easy? Well, take a couple of minutes and try it. How about this?

First, pick up the box of matches and extract one match.
Problem. What happens if you pick up the box upside down – all the matches fall on the floor! Try again.

Pick up the box of matches. Make sure that the box is the right way up. OK, that's better. But there are nasty folk who put the container of matches back upside down in the box so that although it looks fine from the outside you'll still have all those matches on the floor if you just open it too quickly.

Pick up the box of matches carefully. Make sure that the box is the right way up. Hold on! How does one know if it is the right way up?

Pick up the box of matches carefully and the right way up. The label should tell you that it is the right way up. Be very careful as you open the box so that the matches don't fall out on the floor.
Better, now we're getting somewhere!

Take one match and strike it against the side of the box.
Hold on, there are usually a lot of dead matches in most boxes, and what about the safety aspect of striking a match, shouldn't you mention that?

Select one match, the unused ones have a full head, the used ones a thin and blackened head. Hold the match between finger and thumb (which finger?) between index finger and thumb and strike the head of the match against the rough side of the box and away from you. You should do this in case any sparks fly up and into the box.

Good, now we're getting somewhere!

You see even from this rather obvious example just how important it is to have the *correct* procedures and to communicate these clearly enough so that there is no loose expression or ambiguous meaning. Opening up a box of matches isn't a big deal. Setting out the procedures for operating that new cement mixer, or that electronic hoist for heavy loads, certainly is. Get the procedures wrong and it could be someone's life at risk.

We saw on pages 9–10 just how dangerous it is to write in an ambiguous way. You always have to remember your reader and try and put yourself in his or her situation. What may be very clear and very obvious to you may not be to your reader. Consider also that a procedure that looks OK on paper will need to be tried out in practice – remember those matches!

Effective procedures are those then which have been tried and tested. They offer the clearest, most direct and usually the simplest method of doing a job. They are written with health, safety and efficiency in mind.

Just because they are written and in print doesn't mean to say that they can't be developed and improved. As you work with various procedures, keep an eye open for ways of improving them. We don't mean short cuts – these usually lead to unsafe working – but to ways in which, as with our box of matches example, things can be written more clearly and more directly.

Activity

Here is an example of a set of procedures. They are concerned with the setting up of apparatus for gas welding. Read them carefully as though you had to implement them. What would you criticise in the way these are written? What improvements could you make to them?

Apparatus for Gas Welding
Read these carefully before using the gas welding equipment.

(1) The oxygen and acetylene cylinders are chained to the cart.

(2) The oxygen cylinder valve is opened to remove any dust present. This step is repeated with the acetylene valve, a T wrench is used.
(3) The regulators are fixed to their respective cylinders.
(4) These are tightened.
(5) The adjusting screws on each regulator are checked.
(6) The twin hoses are then fixed to the regulators. It is important that the red left hand thread goes to the acetylene regulator and the green right hand thread goes to the oxygen regulator.
(7) Finally, the twin hoses and the welding torch are secured following the same left and right directions.

Well what did you think? Here are some suggested improvements. Would you agree?

(1) Chained to the cart. Should some mention be made of how these are to be chained to the cart? Presumably there is some recommended/approved/required method; if so then that should be stated.
(2) The valve is opened. Should the reader be told how much, i.e. slightly/a lot or, more accurately, by what number of turns on the wheel? Does this mean fully opened?
(3) A T wrench is used. Does this mean that the wrench is used only with the acetylene valve? It could imply that both the oxygen and the acetylene valve need a T wrench!
(4) Tightened? To what level – hand tight? Again this should be stated.
(6/7) On the surface these two read reasonably clearly. The only way to test these is to do the actual operation. It might be that after that we should want to add some detail to these instructions, i.e the same left and right directions. This is rather vague. Will it make 100% sense?

One way to think about procedures is to see the various stages as in a flow diagram or list form. Here is an example.

Scaffold erection

```
┌─────────────────────────────────────────────────┐
│                                                   │
│  Materials for scaffold delivered.                │
│                                                   │
├─────────────────────────────────────────────────┴──┐
│  Materials inspected and signed for      Signed      │
├──────────────────────────────────────────────────┐  ┘
│    Scaffold erected - supervised by               │
│                                                   │
│      Inspected when erected              Signed   │
├─────────────────────────────────────────────────┐└──┐
│        Inspected during use                       │  │
│                                                   │  │
│          Action taken                    Signed   │  │
├─────────────────────────────────────────────────┐└──┘
│        Results of action                 Signed  │
├─────────────────────────────────────────────────┐└──┐
│        Scaffold taken down                        │  │
│                                                   │  │
│          Materials               Signed out       │  │
└──────────────────────────────────────────────────┴──┘
```

Writing it this way makes it easier to see at a glance what has or has not been done, where the gaps are and who has signed off each stage of the process.

If you are planning a new set of procedures we would recommend that you set them out in flow chart fashion. Do this on a large sheet of paper and start off in pencil.

```
1. Mix sand with lime--->|2. Add water--->|3. Mix until a paste--->
────────────────────────────────────────────────────────────────
4. Leave for ten min.--->|5. Pour on to base--->|6. Smooth with spade.
```

The true test is: can someone else by reading through your procedures do the job as it was designed to be done? If not, go back and have another careful look at your flow chart, perhaps you missed out a stage; perhaps you didn't make one stage clear enough; perhaps you assumed too much? Do you think any reader of these procedures would be clear enough as to what you meant by *Mix until a paste*? or *Pour on to base*?

Final point: one of the worst aircraft accidents in history was caused when a baggage handler failed to follow the door locking procedures as they had been laid down. The main baggage door burst open when the aircraft was in flight and this caused a sudden loss of pressure which resulted in the aircraft plunging out of the skies to crash in a wood near Paris with the loss of most of its crew and passengers.

Record keeping

In the construction industry, as in others, it is vital that you get into the habit of making and keeping accurate records. Because firms have all kinds of demands placed on them – from taxation, health and safety, environmental protection, waste disposal etc. – the forms for these need to be clearly completed, dated and signed. There is no future in the odd scrap of paper. The various authorities will not be impressed by a bundle of these.

> Dear Tax Inspector,
> Here is a bundle of bits and pieces which I've collected for you during the year. I hope you'll find them useful! You may also find some odd bits of chewing gum and half eaten crisps – sorry!

Those of you who work for yourself, or want to in the future, may well think that this paper filling is all a chore – but essential forms have to be completed. If they have to be done – and that's non negotiable, it is the law – then they might as well be done correctly.

The temptation in the industry is to rush these forms, get them out of the way as quickly as possible so as to get on the with building, joinery, plumbing, etc. The problem is that if they are rushed out of the way then mistakes happen and even more time has to be found to put them right. So with records, as with so many things, it is a case of Get It Right First Time!

These records also have to be kept in a safe place so that when you are asked to show them you can quickly find them rather than having to search about in the bottom of several drawers and tool cupboards. Murphy's Law states that what you don't want to happen will happen, i.e. the toast always falls on the buttered side, the car always fails to start when the prettiest girl/best looking guy is waiting for you to collect him or her! This law being what it is, you can bet that the form you most need and in the quickest time will be the one that is at the bottom of the greasiest tool drawer!

Let's look first at a very common kind of record, the accident record. Under the Health and Safety at Work Act 1974 all accidents have to be recorded so that:

- the problem can be investigated so as to prevent similar accidents happening in the future;
- there is an accurate record in case of any legal action – the employee against the employer, the employer against the employee or any third party action, i.e. some member of the public, a worker of another company;
- information on all accidents can be collected and used by the authorities to identify trends – i.e. an increasing problem with this kind of machinery, this particular process.

Any accident record should answer these key questions:

When?   Date and time of the accident.
Who?   Names of those involved and any witnesses.
Where? Which part of the site?
What?   Concisely, what did happen? Diagrams may be useful to supplement the written word.
Why?   Are there any reasons for the accident? Any possible causes? Is there a likely sequence of events?

Activity

Read the following situation which happened on site. How would you construct an accident report form using this information? Consider those *when–why* questions. Jot down a few headings to help you.

March 30th 1999. Employee Gary Jones working on building site – Scott Ave Lichfield – laying new paving slabs in front of recently completed houses. Light drizzle – it had been raining hard earlier. At about 3 p.m. he slipped on some wet cement on newly set paving slabs. According to worker he forgot to notice wet cement behind him as he was working to the front. Fell and sprained left wrist. Sprain meant he was unable to do slab laying. In some pain. Saw supervisor who signed him off. Went to first aid box and put on temporary bandage. Attended GP clinic that evening and had wrist re-bandaged. Returned to work two days later, April 2nd, able to resume work, pointing wall. Unable to resume slab laying for further two days.

Activity

Here is the accident report form which was completed after this accident. What do you think of it? What changes if any would you make to it?

Make Haste Builders

Accident On Duty Record

The following accident occurred on site. Details are given below.

Date of Accident
March 30th 1999
Time
Afternoon
Place.
Construction site Scott Ave.

Description of accident:
Employee slipped while laying slabs
Type of injury and part of body affected:
Sprained left wrist.

Names and addresses of witnesses:
Jeff Richards, Supervisor.
What protective clothing/equipment was being used?
Outdoor wear

Signature
J Richards

Job title.
Supervisor
Name (block letters)
J RICHARDS
Date
April 7th

Compare again the information provided and the resulting record. Let's use those when–why questions as pegs for our criticism.

- *When?* Was there enough detail here? It is important to get the detail right. Perhaps in this case it is not that important to have the exact timing, but as far as the Health and Safety at Work officials are concerned, only if the timings are provided do patterns begin to emerge. For instance, is it true that most of X type accidents tend to happen at Y times?
- *Who?* Again is there enough information here? We need more information about the employee. Was he part-time, full-time, part of a contracted out service? Was he an apprentice? Was he an older worker (needing eyesight tests?) who should have known better or have been instructed better?
- *Where?* This is rather vague. We need more detail of where it happened. Simply to report that it was on the site isn't good enough. We need to know which part of the site; whereabouts was the person when the accident happened? Was he working alone, in a gang, under direct supervision? All this should be included in the record; remember this is a legal document. If there's any action in law to be done then we'll need the facts.
- *How/what?* There again it is sparse as far as information is concerned. Fell and sprained his wrist is a very bald statement. It is important that weather and working conditions are noted. Does Gary Jones have many falls? Have other workers suffered falls doing this kind of job? Is there something unsafe in this slab laying practice? Should there have been more supervision?

You might think that this is going over the top – making a right old fuss about a sprained wrist. But think, it could have been a broken wrist – the man's off work for weeks; it could have been a fall resulting in a severe knock on the head, with him in hospital and off work for months! Think of the loss of income for him and his family, the trouble caused to the firm, inconvenience to the customers, etc. Making a full, readable and yet concise record of a 'small accident' like this one may in fact save a great deal of money and inconvenience, never mind future contracts!

- Why? Well this report doesn't tell us anything. Some estimate of the why should be entered, even if it is only very approximate.

Let's see after these criticisms how the report might have been written.

Activity

Read this version. Would you still want to make changes?
If so which?

Make Haste Builders

Accident On Duty Record

The following accident occurred on site. Details are given below.

Date of Accident
March 30th 1999
Time
3 pm
Place
Construction site 12 Scott Ave. Elborough

Description of accident:
G Jones, full time employee (bricklayer) slipped while laying slabs. It had been raining heavily during the morning.
Ground wet. Employee working alone, appears to have slipped
on some wet cement on newly set paving slabs. According to
eye witnesses and in talking with employee it seems that he
simply forgot to notice the pile of wet cement behind him,
was distracted by something and in going backwards to fetch
a spade he slipped and fell. After reporting to supervisor
wrist was bandaged and he was sent home.

Type of injury and part of body affected:
Fell on left side, spraining his wrist.

Names and addresses of witnesses:
Jeff Richards, Supervisor.

What protective clothing/equipment was being used?
Outdoor wear. Heavy boots

Signature

Richards

Job title.
Supervisor on site
Name (block letters)
J RICHARDS
Date April 7th 1999.

Although each company may well have its own accident report form, under the Health and Safety at Work Act 1974 the reporting of accidents has to be carried out on an approved HSE form. This asks the writer to set out the circumstances under

Part A – about you
Part B – about the incident
Part C – about the injured person
Part D – about the injury
Part E – about the kind of injury
Part F – dangerous occurrences
Part G – describing what happened
      – the name of any substance involved
      – the name and type of any machine involved
      – the events that led up to the accident
      – the part played by any people

It is crucial that you follow these actions carefully if you don't want to have the form returned to you.

Other reports

There may be occasions at work when you are asked to write a short report. You may have had to do this at college. Reports should have:

- *An introduction.* Why this report is being written, plus some background information.
- *A middle.* The main points or findings.
- *A conclusion/recommendations.* A summing up plus comments as to what should be done.

Reports should be numbered and clearly set out. Before you start writing you need to ask yourself some important questions.

- *Why* is this report being written?
- *Who* is it for? Architects, quantity surveyors, managers of the company, health and safety officers, other contractors ?
- *What* do they want to know?

When you've answered these questions think about the following stages.

(1) Collect the material – the evidence. This could be in the form of notes, drawings, photographs, other reports, customers' letters, etc.
(2) Sketch out a plan, something along the lines of:

```
Reasons for writing report and introduction
----------------
----------------
Main points (in order of importance)
----------------
----------------
----------------
----------------
Conclusions
----------------
----------------
Recommendations/Action
----------------
----------------
```

Activity

Here are the notes for a short report. Before reading the final version think how you would organise them for a short report. Try writing a rough draft using this information and then compare it with the actual report.

Notes

Report on Present First Aid Arrangements at Bodgers Brothers Builders
(These matters are covered by the Construction Regulations 1966)

Complaints from workers on site, difficult to locate first aid boxes, recent accident at Greenhills site – worker unable to find necessary bandages, etc. – several other problems – replacement first aid materials not being ordered, system breaking down – box at Greenhills not replenished for months.
More than one box needed on bigger sites – record book of use of first aid not kept up to date – at one site Crossfields no book seen for a month – problems when it comes to writing up health & safety reports.
Need big notices round site telling people where first aid box is kept – present notices too small and tend to be hidden – different types of box at different sites – better to replace them in future with same type so that people will know where to find things.

Report

BODGERS BROTHERS

Report on First Aid Provision on Site

Introduction
This report examines various problems with first
aid boxes on sites run by the company. At present
each site has a first aid box. A number of
complaints have been made by workers.

Main points
(1) Difficulty in finding the boxes.
(2) At large sites more than one box is needed.
(3) The first aid boxes are of different types.
(4) Replacement of used first aid materials is not
     working well. One box not replenished for
     months.
(5) The record books which should be placed
     alongside each box are not being kept up to
     date.

Conclusions
The present system is not working well.

Suggested actions
(1) New first aid boxes for each site.
(2) Large notices for each site relating to
     whereabouts of first aid boxes.
(3) One person at each site to be given
     responsibility for replenishment of boxes.
(4) New record book to be bought for Greenhills.

Suggested review of how the system is working
following these changes.

Activity

What do you think of this as a report? How could it be improved?
Jot down your ideas and then compare it with this second version.
The words in italics provide more information from the notes supplied.

BODGERS BROTHERS

Report on First Aid Provision on Site April 3rd 1999

Introduction
This report examines various problems with first aid boxes on sites run by the company. At present each site has a first aid box. A number of complaints have been made by workers *concerning these boxes*.

Main points
(1) Difficulty in finding the boxes. *Notices are not big enough and easy enough to see around the site. They should show staff where these are located.*
(2) At large sites more than one box is needed.
(3) The first aid boxes are of different types. *It is difficult in an emergency for workers to locate the required bandages etc.*
(4) Replacement of used first aid materials is not working well. One box had not been replenished for months. *Following a recent accident to a worker at Greenhills, he was not able to find bandages. Luckily a mate helped him with a clean handkerchief.*
(5) The record books which should be placed alongside each box are not being kept up to date. *At one site the book is missing and has not been seen for over a month. This makes it difficult if not impossible for Health and Safety records to be maintained.*

Conclusions
The present system is not working well. *Company workers are put at some risk. Health and Safety records are not being completed as required.*

Suggested actions
(1) New first aid boxes for each site – *existing ones to be used as back ups.*
(2) Large notices for each site to be displayed relating to whereabouts of first aid boxes.
(3) One person at each site to be given responsibility for replenishment of boxes *when supplies are used*.
(4) New record book to be bought for Greenhills. Other record books to be kept up to date.

Suggested review of how the system is working following *in six months time.*

Do you feel this is a more useful report – useful in the sense that it tells the reader more of the key information that he or she needs to know?

Activity

Here are some questions for you to ask yourself on this and any report you write.

	Yes	No
Is it obvious from the introduction what the report is about?	❐	❐
Is it clear who should be reading this report?	❐	❐
Is the introduction clear and does it put the reader in the picture?	❐	❐
Are the various points clearly set out?	❐	❐
Do the conclusions cover the points in the findings?	❐	❐
Do recommendations cover the points in the conclusions?	❐	❐
Is the report as brief as it is possible to make it?	❐	❐
Is the report clearly hand written or word processed?	❐	❐
Is it free of spelling and punctuation errors?	❐	❐

Almost all of you reading this will be asked to write a report, even just a one page effort, at some time so it is important to get as many ticks in the YES column as possible. Make use of this checklist when you next have to write a report. It might be useful to show it to a colleague who has to write large numbers of reports or one who has never written one before.

Memos

A memo is a short notice sent out to inform people as to some action required, change of procedure, information relating to a visit, inspection, etc. It is supposed to be brief and very easy to read. A memo may be posted up on a noticeboard or delivered to people. It should not be more than one side of paper.

Here is an example of a memo written after the report on first aid boxes had been approved by the management of Bodgers Brothers.

FIRST AID BOXES

Following a survey by Jim Smith and subsequent
report would all company staff please note.

New first aid boxes have been located on each
site.
These boxes are now located in site offices.
When any contents are used, a record card must
be completed.
Record cards can be found next to boxes clearly
marked.
Site first aiders are responsible for
replenishing stock.

Any queries regarding first aid boxes please refer
to site office or to

Company Health & Safety Officer
Harry Grant on 011-222-333 Extension 45

Activity

Is this a readable memo? Will it make sense to workers at Bodgers as their staff pass
by and read it on the noticeboard? Spend a few minutes redrafting it.

You might consider the following.

- Will a phrase such as: *Record cards can be found next to the boxes clearly
  marked* be clear enough? Should some mention be made of how
  these cards should be filled in and what happens to them after that
  – i.e. are they stored, sent somewhere (for instance to the company
  office) or filed in the site office? If nothing is written down then you
  can bet that many of those cards will find themselves on the floor
  and then in the bin.
- How about a date on top of the memo. There's no point in putting
  up a memo on a noticeboard without a date being included. Too
  many noticeboards are never attended to because most if not all the
  various notices are well past their sell-by date.

Basically if the message in a memo is not 100% clear after a first read
then it should be rewritten. People haven't got the time to scratch their
heads or ask questions of others in order to find out what a memo
means. It must be 100% clear at first reading.

If you write a memo then do show it to a couple of people you will want to read it. Ask them if it's clear. Ask them if there's something missing or something that should be added. If it's going to be put up on a noticeboard then do make sure that the print size is big and bold and that you use a drawing pin at each corner. Think of any noticeboard you have at work.

- Is the noticeboard in a place where everyone who has to see it can see it?
- Are the notices readable from a distance or are they in pencil and only designed for those who stand six inches away with reading glasses on?
- When, if ever, do they get updated? Is it just when they fall off, the pins come out or people write rude words all over them?

Personal checklists

These can be issued to individuals or used as information on noticeboards. They serve as reminders of what should be done. Here for example is one concerning health and safety.

Health and Safety Checklist	
Item	☑ **When done**
Safety policy Find out the company's policy and particular aspects which apply to YOU:	❏
Personal responsibilities How you should switch off certain machinery in the case of a fire alarm sounding.	❏
Safety literature Have you read the key leaflets and notices?	❏
Key safety people Find out who they are, what their functions are and where you can get hold of them in an emergency.	❏
Key safety systems Find out about any particular hazards which may be associated with your own work areas. Make yourself aware of safe working practices.	❏
Prohibited areas Find out where they are.	❏

Safety equipment
Find out when and why it must be used and who is
responsible for training and the maintenance of the systems. ☐

Smoking
Observe the company's smoking policy.
If you are a smoker find out where it is safe and permitted
to smoke. ☐

First aid and first aiders
Discover who they are, where they can be found and
how to reach them. ☐

Accident procedures
Find out what to do in the event of a fire, bomb alert.
Find out what the evacuation procedures are. ☐

Checklists are very useful in that the user can see at a glance what it is
he or she is supposed to do.

Site logs/instruction books

This is a very useful way of recording information. It can save a lot of
memos and telephone conversations. Basically it is a tear-out page
with carbon sheets so that duplicate, or even triplicate, copies can be
made. The questions are posed by the foreman in the left hand
column, against which any answers are put down in the right hand
column. These replies can be given by someone who is in overall
charge – a supervisor or a sub-contractor. Here is an example.

Build Em High Construction
Railways St Site Elborough   Job 2345 Page 31

Questions	Answers
24/What is the chosen colour for interior walls in kitchen?	Colour no. 123 light rose
25/When will delivery be made of extractor fan systems?	Week beg. 3 Aug
26/Waste water pipes show signs of corrosion. When can we get replacements?	Noted Report at meeting on Tues 4th Aug.
27/Please confirm removal of debris near site entrance	Sub-contractor promises this Frid

With such a book it is vital that the supervising officer signs each item and that the reply is dealt with on each visit made. When a whole page is completed he or she should take one copy for the records, one copy should be sent to the contractor's office and one copy should be kept on site. If quantity surveyors are employed on the contract this copy should be shown to them on their visits to assist them in keeping in touch with what is happening.

Progress charts

This is as it says a means of displaying where the job has got to in relation to where it should be by a certain date. Such a chart should be easy to see at a glance. The use of colour can be helpful. We shall use black and grey, but it will give you the idea.

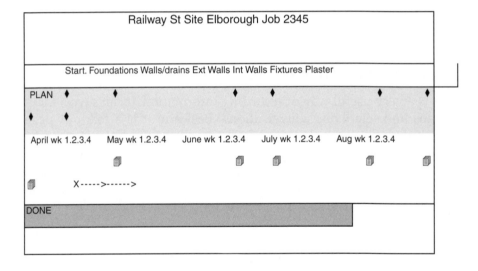

There are all kinds of progress charts. The point is can you quickly see where you've got to as opposed to where you thought you'd be! In this one we can see that progress is matching the plan well but the plastering is falling behind and won't be finished until the second week of August as opposed to the planned week 4 in July.

Meetings and minutes

Where there are problems with the progress of a job then site meetings are usually called. You may never have been to a meeting but we can guarantee you that you will! Most people's views of meetings are that they are boring, a waste of time and a chance for windbags to have their say. You may be attending a number of informal meetings as companies increasingly involve employees in decision making through team working.

If you become a supervisor then you'll find yourself involved in many meetings. If you start your own business you may well be involved in meetings with contractors, quantity surveys, architects, etc. These meetings may be to do with trades unions and working groups; and many will be one-to-one meetings with colleagues and managers. Some of the meetings with site foremen, architects, etc. will be stand up affairs involving you for a couple of minutes. e.g.

Foreman	OK I wanted to have this quick meeting to look at deadlines. As you know we're due to complete all the inner walls by October 1st. Any thoughts?
Jim, Supervisor	Should be OK but there's been a delay in the delivery of stone from quarries for the outside walls.
Foreman	What's your estimate for completion?
Jim	Well assuming we get delivery next week, then Friday week.
Foreman	OK well let me know if there are any more delays. Any one else with any problems?
Harry, Supervisor	There's been some flooding after last night's rain but it should be dry enough later this morning for laying drainage pipes.
Foreman	OK keep me informed if you can't complete today. Any other problems? OK next meeting here 8.30 am Monday.

Some meetings are much, much longer than this. You won't be able to escape! So given this fact here are a few tips to assist you.

- *Do know why you're attending the meeting?* There's absolutely no point in just turning up hoping that everything will become clear when you sit down at the table.
- *Try and find out what's on the agenda.* The agenda is a list of those topics to be discussed before the meeting, giving you a chance to prepare. You might want to bring some papers or evidence; you might want to talk through the agenda items with your colleagues, especially if you represent them in any way.

- *Check your timing.* Warning! Meetings take up time. There's no point agreeing to go to a meeting unless you find out from the organiser, convenor, chairperson how long it will last or, better still, when your particular bit will be discussed – that means that you may not have to be there all the time.

A typical agenda

Agendas come in all shapes and sizes but this is a fairly standard format.

```
                    BODGERS BROTHERS

                    Training Group

The next meeting of the group will be held on
Wednesday Oct. 5th 1999 at 2.30 pm main office 12
Shore Rd Elborough.

AGENDA

  1. Apologies
  2. Minutes of last meeting
  3. Matters arising
  4. Feedback from college tutors
  5. Training awards
  6. Discussion of on-site supervision of students
  7. Any other business
  8. Date of next meeting
```

Agendas should be sent out to persons attending the meeting several days in advance. It is important for the 'health' of the meeting that you should be able to place the items of business that you and those whom you represent feel should be discussed on the agenda.

It helps to understand the particular language of meetings. Too many people sit at meetings not able to say much because they haven't got a clue what is going on – it's all a foreign language as far as they're concerned! Let's explain the language of agendas.

- Apologies. These are explanations sent in advance when someone can't attend the meeting. It's a form of politeness. Furthermore, if you are representing a group – workers from a particular site for instance – and you seldom if ever bother to turn up for the meetings, then that fact should be known. The site group deserve to have their views represented. After so many absences it may be time to find another rep!

- Minutes from last meeting. We'll be looking at these on pages 87–9. Basically they are brief notes on what was discussed under each item on the agenda and what, if any, decision was made?
- Matters arising. Any outstanding items of business which will not be covered in the main agenda.
- Any other business. Any extra items for discussion. If your group only meets infrequently then it will be important to leave time to discuss any items which have come up since the agenda was fixed.
- Date of next meeting. Be careful! It sounds obvious but do think very carefully if you have to attend this next meeting – it may be that your presence is only needed for certain ones or as we said earlier only for a certain part of the meeting.

At the meeting

We've looked at an informal meeting, you and a couple of others talking through a particular problem – the site meeting. You will occasionally come across more formal ones. These often have rules and procedures. People sit round a table as opposed to standing in the mud in the middle of a site or squatting on tables in a site office.

Here is a guide to some of the language used in these formal meetings. You might as well know these terms. You might need them if you ever become a committee member of chairperson for a union, sports club, community association, parent teachers' association, etc.

- Quorum. Someone might say. '*Is this a quorum*?' or '*Are we quorate*?' This means are there enough people at the meeting for the business

to continue? Some organisations have very rigid rules about minimum numbers. The argument for this is that if the meeting is all about representing various groups' views then enough representatives ought to be there. If only three out of a possible 15 turn up then you can't really have a meeting, you're not a quorum.

- Speak through the chair. It's an odd expression; what it means is that you are invited to speak by the chairman and all your remarks have to be made though him or her. In other words you can't start a conversation with your neighbour or shout 'Rubbish!' across the room at the person who's speaking.
- Who'll propose this? There's been some discussion on a topic and someone's suggested some action, some way out of the log jam. The chair will ask this question hoping that there is someone round the table who will put their name to this action. It can then be seconded (agreed to by another person) and then accepted or put to the vote.

There may be other words and phrases used during the meeting. Listen carefully and you'll soon learn the language. Ask your neighbour; he or she may know, on the other hand perhaps they've been sitting in baffled silence for the last dozen meetings wondering what's going on.

Your role at the meeting

Basically you may have one of three roles.

The technical
This refers to the particular expertise you bring with you as a builder, joiner, plumber, etc. This technical expertise allows you to speak when it comes to any 'technical' issues that may be discussed. e.g.

> *Harry, any hold ups as far as plumbing is concerned?*
> Well there could be problems with completing the connection
> to the surface water drains. Let me show you on this plan.

The representational
This could be a trade union role or a health and safety role, etc. in which you are acting on behalf of others, whether in an official capacity (you've been elected) or an unofficial one (someone's just asked you).

> *Harry anything you'd like to raise?*
> There's grumbles from the guys about overtime payments – some
> of them have not been paid and this is for work done two months
> ago!

The unique
This is what you bring because you are yourself – your 'unique' personality.

A researcher called Belbin has spend many years investigating what individual roles people can bring to groups. Here is a summary of the types he has 'located'

- **Shaper or pusher.** Impatient; keen to get on with it. Likes to 'push' the group along. Uses such phrases as:

  > Right so what are we going to do about it?
  > So what's been decided then?

  This is a very useful person to have in any team and particularly useful at a meeting where people like to hear the sound of their own voices! They're useful when the clock is ticking away and some decision just has to be made. On the down side these people can get a bit impatient and can be fairly abrupt. Perhaps that's the price we have to pay for having them around.

- **Innovator, ideas person.** Good at producing ideas and solutions; sometimes a bit off-beam, sometimes with their head in the clouds:

> Well we could do it this way.
> Why don't we turn the whole thing round?
> How about this for a solution?

These people can be so useful at any meeting. They bring in fresh ideas, new ways of looking at a particular problem. The down side is that very often their heads are in the clouds and they're often not paying attention to what is being said – they're still thinking about what was happening 20 minutes ago!

- **The completer.** The person who likes to see things through to the end, who likes tying up loose ends, who's good at detail.

  > We'll need to send that back before Thursday.
  > Who's going to book the room then?
  > We'll need to phone first and check.

  These people are incredibly useful – essential even – for the health of the team. They're the ones who see to the detail. It's all very well having great ideas but we also need someone who will make sure that the t's are crossed and the i's are dotted. The down side to having them on board is that they often worry over-much about these small details and so can become a real pain in the neck.

- **The evaluator.** The one who is good at asking questions where others forget to; the one who is good at spotting 'holes' which others have overlooked.

  > What happens if we can't get permission for this?
  > What would the legal position be?
  > Do we have anything on paper to cover this?

  These are very useful people; they often ask the questions we wish we had. They're good at finding holes in the argument; they'd make good lawyers and even better detectives. The down side to having them around is that they can hold up proceedings with all these questions.

Belbin has eight particular roles which he thinks should be included if there is to be a successful team. He is not saying that unless you have these eight roles you will not have a good team. Rather, he suggests that these types provide a mix which will help make your team and your meetings more successful.

The important thing from your point of view is to play the role you feel most happy with and encourage others to do the same. It's not a question of play acting, of pretending to take on a role that you don't feel is you (today I'm going to be the ideas person – I'll put on my baseball cap with GENIUS stamped on it!). It's all about making sure that you feel confident to play the person you really are and able to contribute freely to any discussion.

Minutes

These, as we've said, are the notes of what was discussed – and hopefully agreed on at the meeting. They are written in the past not the present. For example, 'He reported that fire alarms were now working', rather than the actual words used, 'The fire alarms are now working.'

Activity

Here are the actual words used in this meeting by Harry Jones who went to Elborough College to meet lecturers in the Department of Building.
Can you write these as a minute?

Chairman, I visited Elborough College and had a meeting with some lecturers from the Department of Building. I expressed our concerns regarding the failure rates of students at NVQ levels. They told me that a report was being put together by the department on this problem. One of them would attend our meeting on November 15th to discuss this. The department would send copies of this report before this meeting so we could all have a chance to read it.

Here is an example of the minutes which followed the agenda we saw on page 82. Read them carefully, especially the passage of Harry Jones' report on his visit to Elborough College. We've underlined the changes from present tense (now) to the past reported tense used in minutes (then).

BODGER BROTHERS

Training Group

Minutes of the meeting held in Room 211 on Wednesday Oct 5th 1999

Present. Alex Smith (Chair). Harry Jones, Pete Smith, Jennifer Armstrong, Jack Tait, Fred Ricak, Susan Snaith, Jim Deal.

1. *Apologies.* These were received from Tom Hope.

2. *Minutes of last meeting.* There were no amendments. Signed by chair as true record.

3. *Matters arising.* There were none.

4. *Feedback on college training.* Harry Jones reported on his visit to Elborough College where he had met with lecturers from the Department of Building in view of concerns expressed at recent failure rates at NVQ levels. A report on these failures was being compiled by the Department of Building. A representative from this department would attend the next meeting of the training group on Nov 15. The report would be circulated in advance so that everyone could have an opportunity to read it.

5. *Training Awards.* The group accepted a recommendation from the chair that an award be presented to the most promising new entrant to the company. This award would consist of a cheque for £500.

6. *On-site supervision.* The group approved the new on-site supervision system as laid out in minute 3 of the Aug. 15th meeting. Susan Snaith was congratulated for her efforts on developing this system.

7. *AOB.* None.

8. *Date of next meeting.* November 15th 1999.

Activity

How readable did you find these minutes? Can you think of any improvements? Are there any other changes you would make?

How about an ACTION column where we list who does what? e.g.

```
                                              Action
4. Training Awards. The
   group accepted a
   recommendation from
   the chair that an award
   be presented to the most     Pete Smith
   promising new entrant
   to the company. This
   award would consist of
   a cheque for £500.
```

Being an effective participant at meetings

Many people are worried by meetings. They sit there in silence, are afraid to ask questions and just wish the clock could move faster to the close of the meeting. But if you have decided to spend your time at a meeting and if you are representing the views of others then it's a shame not to ask questions, offer suggestions and generally participate.

One of the best ways to participate in any meeting is to ask questions. Here are some approaches:

- Ask if you don't understand a term, an abbreviation, etc. A lot of words are exchanged at meetings and many of those sitting round the table haven't got a clue what is being said. So do ask: 'Chairman, can I ask does ... that mean? Excuse me. Can I have a translation for that please'.
- Ask a question if you want to know more about something being discussed. 'Can I ask when the award will be paid?'
- Try to exchange ideas with others round the table. 'Does anyone else agree with this view?'

The important thing is to try and get to your meeting early so that you can meet up with the others before the formal business starts; this way it will be easier when you come to clear your throat and ask questions or make a point during the proceedings.

Making a case at a meeting, a short presentation

There will come a time when you are asked to present a plan, a course
of action or a way of making improvements in your work to a group
of people who will be sitting round a table watching you and
hopefully listening to you. Yes, it's all rather like an interview and
remember in our section on being interviewed (pages 51–4) we looked
at how to give the best possible impression of yourself. The same
advice holds good for presentations to a committee.

- Find out what is expected of you – the level of detail for instance.
- Find out how long you're supposed to speak for.
- Find out if there are any visual aids such as a chalk board, white-
  board, flip chart, video player, etc. They might be useful to illustrate
  your presentation.
- Find out who'll be there. Are they architects, planners, builders,
  inspectors? What is their level of knowledge, their experience?
- Keep it brief and to the point – don't waffle on.
- Jot down a few points that you wish to make. Write them nice and
  big and bold on a card – big and bold enough so that you can see
  them from a distance. You don't want to have to peer at them; that
  will only make you feel even more nervous. Make your list of key
  points like this.

> 1 Intro
>
> 2 Cause of problem
>
> 3 Solutions
>
>   • ....
>
>   • ....
>
> 4 Actions
>
>   • ....

But not like this.

> One of the main causes of
> corrosion found in the
> joints was due to the
> wrong setting being used
> to tighten the various
> bolts. The results of this
> undermine the stress
> from above and eventu-
> ally cause very real
> difficulties for builders.
> Safety here . . .

- Try to look at the members of the group and don't look down at the table. That's why it's important for you to have your notes in an easy to read form.
- As you prepare your presentation think of the possible questions that members of the committee might ask you. Think what you'd ask yourself if you were listening to your talk!
- Sometimes it will help you (steady your nerves) and help the listeners if you have something for them to look at – a chart, a model of some kind, picture, a bit of evidence such as a corroded joint, etc. We're not suggesting any fancy visual aids; just something which will reinforce your words.
- If you do use models, slides, large scale photographs, etc. then do check your sight lines. There's no point going to the trouble of getting such visual aids ready if your audience can't see them properly.

We could say a lot more about meetings and your part in them but these things are very well covered in other books. The crucial advice is:

- Do check on the reason for the meeting. 'Excuse me. Before we start can I just check that this is the site safety meeting?'
- Check your role in it. 'I'm representing workers at the Railway St site. They asked me to raise . . .'
- Don't allow your time to be wasted. 'Excuse me. I need to leave before 2 pm. Can I give my report soon please.'
- Be prepared to contribute to the discussion. 'Can I add to that we've seen several examples where first aiders . . .'
- Have the courage to ask if you don't understand. 'Could anyone explain what IIP means. Oh, Investors in People . . . fine!'

Now that we have mentioned asking questions in a meeting we turn to another situation where it is vital that you do plenty of asking.

Being supervised

Most of you reading this are being supervised by someone at work. If you're at college then you have college lecturers supervising you. If you are at work on placement then hopefully someone who is experienced will be helping you and keeping an eye on you at work.

The advice we gave when attending meetings – ask questions – applies equally to being supervised. Any supervisor can only do so much. Few of them are mind readers. If you don't understand something they ask you to do then say so. It's much better that you ask at the start of any job and get it clear rather than run the risk of staying confused and making a mess of it, possibly causing injury to yourself and others.

Just think for a moment of two likely reactions of any supervisor to you, a newcomer to the site, the shop, the crew.

Here's another of those youngsters, wanting my help I suppose. Wasting my time. I do wish they'd go to someone else.

Or

Here's a youngster. Bound to be difficult for him/her. I remember myself what it was like starting off. Well I'll do my best to help.

We hope you find more of the second than the first kind of supervisor. If you are unlucky and find the first type then remember his or her negative view of the job of being your supervisor could be due to the fact that they were only given the job a few minutes beforehand and were actually in the middle of trying to finish something else. It may not be you personally who is the source of annoyance but the lack of communication from their supervisor. We suggest that should you encounter these negative attitudes you are quite open with your supervisor about the situation. Be up front with him or her. e.g.

Hello my name is Jim. I'm doing a placement from college as part of the ... course. I was asked to see you. I guess you're my supervisor for the next couple of weeks. Sorry about that. Expect you're busy enough. I'll try not to waste your time.

This kind of approach followed by a few direct questions as to what you're supposed to do will have much better effect than

Hello. I'm from college. I was told to see you. You're supervising me this week. What do I do?

Here are some tips which, if they don't make your supervisor exactly smile, should at least make him or her less likely to be negative towards you.

- Try and think through some questions that you need to have clarified before you meet the person who is to supervise you.
- Take a small notebook and a pencil and then you can jot down any particular points he or she will want you to remember. From a supervisor's point of view one of the most annoying things is to have some youngster keep asking questions to which they have already supplied the answer:

    Excuse me did you say a couple of minutes ago move those bricks . . . but sorry I've clean forgotten . . . where to?

- Be sensitive to your supervisor's needs. He or she will probably have many other things to do. If you have to meet with him or her try and 'negotiate' a good time. Don't just assume that they will be prepared to drop everything because of you and your needs.
- Make sure you get some feedback on your performance; it's your right. Ask for it in a reasonable way. Try and negotiate a good time for both of you so that your supervisor can give useful advice and point up some of your strengths and areas to work on.
- Don't borrow their tea mug, help yourself to their biscuits, sit in their seat or take their newspapers! Observe what's going on and do not assume just because something's left lying around that it is OK for you to use, eat, drink from, read, wear, write with it, etc. Be particularly sensitive to other people's possessions, especially tools.

One certain way for supervisees to become very unpopular very fast with their supervisor is to (a) 'borrow' tools in the first place, and (b) not return them.

Supervising others

Sooner or later you will be asked to supervise another person at work. This may be a student from college on work placement or a junior starting off.

Activity

Think back to a situation where you were supervised. What was the experience like? If you were to supervise someone what would you try and do differently/the same as when you experienced it ?

Here are some key areas when supervising others.

- Communicate clearly what it is you want done or expect should be done. It's not much help to the person you are supervising if you simply say *'Just get on with that OK'*. You need to be able to provide more definite information.
- Check back with the person you are supervising to make sure that he or she understands what it is you've asked them to do. There's not much point in just saying, *'Have you got that?'* Remember how you felt in situations like that? You were probably not very keen to admit that you didn't understand. So do some checking back.
- If possible jot down the key points on paper. This will help the other person remember any exact points or the sequence the work has to be completed in.
- Try if possible to identify any problems that might arise. Most jobs have their tricky bits; go over these and where possible provide clear advice as to how to deal with them.
- Remember to separate out those parts of any job where the person you are supervising can use his or her initiative and those parts where because of health and safety considerations, etc. the job must be done in a particular way.
- Do give some kind of feedback on the work of the person you are supervising. Avoid just saying, *'That's OK'* or *'Fine'*. He or she will need to know where the job was fine and where it can be done better. Your responsibility is to give useful feedback. If a job's been done well then say so and point out which features have been particularly successful.

Giving praise is the easy part of any supervisor's job. The real challenge is to provide constructive criticism so that the other person actually learns from the experience and will do the job better next time.

Activity

Think of times when you have been criticised by a supervisor at work or lecturer at college. Was the criticism constructive – in other words did you learn from it? Was it so painful that you got angry and ignored the criticism? Read these two examples of a supervisor giving criticism. Which one do you think could be called constructive? Can you spot the differences between the two approaches?

Example 1

You finished all I asked you? Right. Come over here. Let's have a look. So you think you've finished this do you? Well let me point out to you what you've failed to do. You've not put in enough nails here – this is too weak. The joints here are not fitted in well enough. I don't like this. That wood there needs much more work on it. You'll get splintering down here where the door swings open. See here, that's not good enough nor that. That's OK and that but this bit will need more preparation. I don't like this here. You can do better than this.

Example 2

OK you've finished. Let's have a look. Considering this is the first complete door you've done there is some good work. Here for instance you've prepared this part well and you can get the varnish on it later. Good, that's a fine finish. Now let me show you where you can improve. Basically the joints here won't do. Can you see why? Have a close look. Yes just there. When we hang the door what's going to happen? That's right there's a danger that it will splinter. So what do you have to do to prevent that? Good. So that's the first thing I would like you to do. Now here you'll need to get a better finish – more sanding down – feel it – it's not nearly smooth enough. OK.

In example 1, the supervisor criticised in personal terms – *'I don't like'*. Try and avoid slipping into the Critical Parent voice – remember that tends to trigger off the Child. Being a supervisor is not about expressing your personal likes or dislikes; it's about the standards required for the job. It's not whether *I* like this particular joint or bit of bricklaying but whether the joint will stand the strain or the wall won't fall down at the first push.

- So, don't be personal in your criticism. Be specific as to the job and the standards required. These standards are increasingly set down by the various authorities. Refer to them when making criticisms. It will allow you to be much more the Adult in your communication with the supervisee.
- Point out the positives before you go for the negatives. Give praise where praise is due. If you start with the negatives the person you are supervising will probably stop listening after a time. Remember a key point in supervising someone is to enable him or her to do the job without you having to watch them all the time. Your job is to help them get better at their work, to learn from their mistakes and to see where they have got it right.

Many of you will remember what it was like taking lessons from the driving instructor. If it's all criticism over what you didn't do, forgot to do, missed doing and should have done, then you'll feel like jumping out of the car and stumping back home. The instructor's job is to help you pass the test so that you can drive safely and unsupervised.

When you become a supervisor, or if you now supervise others, the test is: if your supervisees can work well on their own without you having to stand and watch them, then you have done your job well. Remember though they may still need your advice, so don't ignore them. Don't dump them. Keep your eyes open. Encourage them to come and ask if they have problems. A two minute explanation is better than having something really serious happen. Remember you will be held responsible in any court of law. You can't blame the supervisee. It was your job to supervise. So keep your eyes open and do a little checking around with your colleagues.

How's that lad from college getting on with you?

Now we need to turn our attention from communicating with those we work with to communicating with those who pay us – our customers.

Group exercise

Think back to when you worked on a site or to any holiday/part-time jobs you've had. Think about any aspect of 'on-site' communication – a report, series of instructions, notices, etc. In your groups consider:

who was the communication designed for?
how was it put into place?
what , if any, were the results?
if the results were positive why was this?
if the results were negative – i.e the communication was ignored – why was that?

5

Communicating with your customer

Customers are important; vital and essential. Our communication with them, how we speak to them, write to them and talk over the phone to them, will be crucial as to whether they:

- place their business with us
- place repeat business in our way
- recommend us to others.

Let's face it, word of mouth recommendation is so very powerful. You know the sort of thing:

> Jim, I know of a good builder/plumber/joiner. Did some great work for me and reasonable charges. Why not give him a phone. I've got his number at home.

That's the kind of recommendation that no amount of advertising can match. It's personal, direct and it has an impact.

Activity

Think of when you have contacted a company – a quote for car insurance for instance or booking a holiday. Did that person representing the company create a positive or negative image for you?

When he or she said, *Can I help you?* did it sound like a genuine invitation, a sincere response to your enquiry; was it spoken with a friendly tone and in a manner that made you feel pleased you had phoned or visited this firm? Did it make you want to do business with them? Or was it more of an unfriendly bark, like a stranger's dog warning you off the premises?

It is so much easier, less expensive and takes less time to keep existing customers than to have to go out and find new ones. Ask anyone in business and they will tell you just how hard it is to attract new custom and why it is so important to keep your existing ones happy. Customer care should be much more than just slogans; it must be part and parcel of one's everyday working life.

Some of you reading this will be fairly new in your workplace, just starting off in a firm while doing your college course or working on a short-term basis, so you may not see customer care as your main priority.

For you that constantly ringing telephone over there represents not so much a valuable customer but a damn nuisance; that person who can't seem to make up their mind represents not a potentially valuable customer requiring help but less time for that well-earned tea break, and that letter of enquiry means just more work. Customer Care! That's all very well for other staff; they can often see the benefits in terms of pay, bonuses, a more secure job, etc.

This is an understandable point of view but we all have to remember that without customers of the right number and right sort (i.e. those who will actually pay) the business will dry up and so will our jobs, especially if we are on the bottom rungs of the ladder or paid on a short-term basis. Despite our feelings, tiredness, need for a cup of tea, etc. customers do have to come first. Customer care has just got to be taken seriously.

Some people don't like the phrase customer care; it sounds a typical business buzz word. Good companies (successful ones) have always gone in for customer care; they may not have called it that but their whole way of working from the time of meeting the customer to the sending of the final invoice is all about customer care – looking after those people who pay the money to keep them in business.

We'll have a look now at various forms of promotion and advertising. We realise that for some of you this is more the concern of your managers and bosses. But the day will come when you will have to think about these things – maybe within a firm and certainly if you decide to start up in your own business.

Some key factors in communicating with customers

Know your customers.

This is the first and foremost aspect of customer care. The more you know about your customers' needs then the more you'll be able to respond to them – even anticipate them. This is why it is so important to build up information on your customers. You can use something as simple as a card index on which you record:

<u>Name of customer</u>

Address
Tel
Fax
e-mail

<u>Type of business</u>
<u>Scale of business</u>
<u>Previous work done for customer by us</u>
<u>Notes on the work</u>

<u>Rates charged</u>
<u>Payment record</u>
<u>Last contact</u>
<u>Response</u>

<u>Likelihood of further business</u>

Activity

Have a look at this customer record card. Read it carefully and then think what you would change about the way the information has been put down.

Name of Customer Build Em Hard Suppliers

Address 14 High St Elborough
Tel   0111-334-557
Fax   0111-339-575
E-mail   Const@em.hardco.uk

Type of business Builders Suppliers
Scale of business Serves Elborough area
Previous work done for customer by us Advised them on
various types of tiles. Tested them.
Notes on the work Tests & meetings after

Rates charged £500
Payment record √
Last contact 1998
Response Might be asked to do some more work

Likelihood of further business Keep in touch.

Did you spot these doubtful areas?

- Types of business *Builders Suppliers.* Is that enough information? Should the record card be a little more specific? What size of firm? Are there any particular specialisms we should know about?
- Notes on the work? *Tests & meetings after?* What will that mean to your colleagues. What will happen when someone from Build Em Hard Suppliers phones up with a request for more help with testing and gets a response 'Testing, dunno what you mean mate'?
- Payment record √ Will that be enough? Does that mean that they paid within 28 working days or was it only after a couple of letters that they finally paid up after three months? Did you have to send in the heavies?
- Last contact *1998* That's too vague, at least mention the month.
- Response *Might be asked to do some more work.* On what – the tiles or some new materials?
- Likelihood of further business *Keep in touch?* Did they suggest that or was it one of your mates? Were there any hints from them which should have been written down, or did the phrase '*Keep in touch*' just mean '*good-bye, farewell, it's been so nice to know you*'?

Such a simple card can of course be expanded. The information can be stored on a computer. The key thing, whether making use of paper or disk, is to *update* the information so that it is an accurate record. Think of just how frustrating it would be when you start phoning a customer, then you realise that one of your mates hasn't bothered to put down that change of address or new fax number.

It is equally frustrating when the last contact from your firm did not write down on the card that the business you're now desperately trying to reach has changed ownership. They're now not making electrical switching gear but are into mobile phone repairs! You're probably the mug who gets the not very amused customer response:

> I told one of your firm last month that we won't be ordering any more. Put that in your records and please don't phone again.

The message is clear: don't neglect customer records. We'll be looking at following up leads from your customers on pages 113–14 but remember, information about customers is the lifeblood of any firm. If it dries up so will your business. Keep the flow moving. Go through your records on a regular basis, weed out the 'dead' ones, adding any information you get from local newspapers, from friends and contacts, local newspapers, etc. Use those bits of information you hear in the pub:

> X company is buying up land near Y and it's likely they'll be looking for some contractors to . . . .

Don't believe all you hear in pubs and about the place, but there's an enormous amount of useful information floating around just waiting to be picked up. Keep your eyes and ears open and don't forget to keep those records up to date. Remember that as you add to them they will become increasingly valuable. Keep copies, back up any files from your computer on to floppy disk. Don't, whatever you do, put all your customer 'eggs' in the one basket – you don't want to come back one day and find them all fried!

Let them know about you and what your business is.
We've already said that word of mouth is perhaps the most powerful form of advertising. However, your organisation, or you as a self employed person, will probably need to supplement that with some kind of promotion. The list is long and your choice will depend on your budget.

A word of warning: there's no point in doing a lot of promotion and looking for new business if, when it comes, you are not in a position to satisfy those demands. You've got to plan for expansion – staff, equipment, transport, computer systems, etc.

Here are a few ways of communicating with your customers and potential customers.

Business cards and stationery

These are such an obvious and comparatively cheap way of advertising your services that they should be considered one of your first forms of promotion. By stationery we mean letter-headed paper – which you will use to send out your invoices on. Handing out a business card as we mentioned on page 7 is so easy and can make all the difference between someone remembering who you are and the work you do and just forgetting you and your firm the moment the meeting is over.

A word of advice. When having your business cards/stationery printed, don't make the print on them too small, e.g.

```
┌──────────────────────────────────────────────┐
│                                                │
│           Port Hall Builders (Est. 1934)       │
│                                                │
│                                                │
│         Brickwork, slabbing, monoblocks        │
│                                                │
│                 Free estimates                 │
│                                                │
│                          Tel 011-222-444       │
│                                                │
└──────────────────────────────────────────────┘
```

This may have looked very neat when the printer set it up on the computer but it is not so helpful when your potential customer is trying to read it while using the telephone or peer at it while trying to get something done in a hurry. This one is much easier to read.

```
┌──────────────────────────────────────────────┐
│                                                │
│        Port Hall Builders (Est. 1934)          │
│                                                │
│        Brickwork, slabbing, monoblocks         │
│                                                │
│             Free estimates                     │
│                                                │
│                    Tel 011-222-444             │
│                                                │
└──────────────────────────────────────────────┘
```

Putting up cards on shop windows

This method is used by large numbers of self employed builders, joiners, plumbers, etc. and much less so by established companies. Putting your advert in the shop window tends to imply that you will not be charging very much for your services.

One of the problems with this kind of display is that the other adverts often look tatty. Do make sure that yours looks smart. If you can't get hold of a word processor or electric typewriter then find a friend who has one. The finished article – a good quality card, well typed and neat looking – will stand out from the hastily written efforts (with spelling mistakes) that so often crowd the windows of corner shops, etc.

As with business cards and stationery, don't make the print size for your name and Tel/Fax numbers too small. A number of more elderly passers-by will be reading your card and will want to jot down the key details so make them BIG and BOLD.

Targeted leaflet posting

This is where your firm goes round selected houses/streets leafleting and you then follow up by calling on those homes where you think there is some chance of getting business. Sometimes there may be give-away signs that some form of construction work is needed – evidence of slates off roofs, broken window frames, rusty guttering, etc.

The problem is that some householders will not take kindly to having these tell-tale signs pointed out; some will already have contacted a firm; and others will never answer the door or simply throw your beautiful art worked leaflets in the bin. You can find distributors who will do the foot slog for you but of course they will charge for their services.

A carefully targeted campaign can produce results but it might just be worth your while looking at other forms of promotion.

Cold phone calling

You know the sort of thing; the telephone rings when you're watching your favourite TV programme or have just started your evening meal and someone wants to sell you double glazing or a new kitchen. Just thinking of this will probably put you off ever doing it yourself. It is a risky way of trying to get business – those that you don't annoy will probably not want to think of buying something over the phone.

Our advice would be to think of some other way of selling; that is unless you have carried out some very specific homework on the type of customer that might buy and where they are likely to live.

Customer surveys

There are various ways of going about this. At the end of a job you can ask your customers what they thought of your service. A simple tick sheet may be enough. You could go in for something more substantial – sending off a survey to a sample of customers you've dealt with over the past year or so. You have to be careful though: there are so many surveys around just now that people are getting fed up with them. Go to almost any café, department store, travel agent and there's a feedback sheet:

> Dear Customer
> Have you got a moment? We want to make sure that our service is the best. Please take a few minutes to fill this in.

Many of these surveys are a genuine attempt to gain information from the customer as to how the service can be improved. You as supplier are also trying to sell, you're finding out where the service/s could be developed and seeking information as to where the gaps are. These gaps may provide you with a chance to develop a new service or one slightly different from others on the market.

Many suppliers try and establish telephone contact with customers following a survey; this provides them with an opportunity to extract more information and to do a bit of selling of their services.

> Thank you Mrs Smith for completing the survey. That was very useful for us. I see you might want to replace a bath with a shower. We can do all the plumbing for you and if you decided to go ahead within three months we could offer you a 10% discount on our normal charges.

You can imagine the risks with this kind of promotion. Your follow up phone call might so irritate the customer that you lose him or her for ever. It has to be done with care.

Information on the firm's vans and cars

These are usually parked where many people are passing by. There's always a chance that someone needing some service in their home will be attracted and contact one of the workers direct or phone the firm. It pays to have any firm's vehicle looking smart and clean. Few will be

attracted by some dirty looking rusting van, especially if this is badly parked so that the whole road is blocked! If the van is driven at high speed, has to swerve to miss pedestrians, if the driver makes rude signs to other motorists and pedestrians and if, as we've said, the van is parked in such a way that it blocks other traffic and people's drive ways then your firm certainly won't be winning any new converts!

Information on boards, displays, etc. set up at the site where you are working

This is an easy to use and cheap form of promotion. Do make sure that the sign looks smart (clear up any graffiti) and is large enough to be seen from a distance and doesn't get in the way of passers-by, shop keepers, traffic wardens, etc!

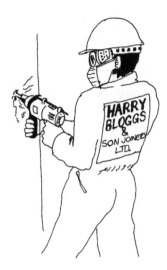

Information on firm's uniforms, hard hats, overalls, etc.

As you and your mates go about your work members of the public will see and hear you. You might go into the local shops for food and newspapers, the post office for stamps, etc. The impression that is made can be positive or negative. Recently we witnessed some workers from a local building company come into a local shop. They shouted for service, were very rude to the staff and obviously made a very poor impression on a number of locals. You can bet that this firm won't be on some potential customers' list of builders to call.

Local shopping guides

These have the advantage of being local and reasonably cheap to advertise in. A lot of people will browse through these. If you're selling joinery, plumbing or building services to the general public then these guides are certainly well worth considering. Here's an example of the layout for such a advert (to scale).

HARRY BLOGGS & SON

JOINERS LTD

PLASTERERS – BUILDERS – ROOFERS

DRY ROT – WET ROT – DAMPNESS

WOODWORM TREATMENT

FREE ESTIMATES

25 YEAR GUARANTEE AVAILABLE

18 Hunter Rd ELBOROUGH

Tel/Fax 011-333-555

It certainly packs in a good deal of information in a short 6.5 x 4.5 cm space. You are obviously restricted as to what you can say about yourself by the small scale. There is no need, however, to have all the printing the same in terms of style and layout. With a little bit of imagination you can make the same information slightly more interesting and eye catching. e.g.

HARRY BLOGGS & SON

JOINERS LTD

PLASTERERS – BUILDERS – ROOFERS

DRY ROT – WET ROT – DAMPNESS

WOODWORM TREATMENT

FREE ESTIMATES

25 YEAR GUARANTEE AVAILABLE

18 Hunter Rd ELBOROUGH

Tel/Fax 011-333-555

A word of advice. Don't overdo the artistic element, like this example for instance!

•••   **HARRY BLOGGS & SON**   •••

JOINERS LTD

PLASTERERS – BUILDERS <ROOFERS>

*DRY ROT – *WET ROT – *DAMPNESS*

WOODWORM TREATMENT

££ FREE ESTIMATES

25 YEAR GUARANTEE AVAILABLE

18 Hunter Rd ELBOROUGH Tel/Fax 011-333-555

☎ **Tel/Fax 011-333-555**

This might be fun to create on the computer but it runs the risk of being so busy and over-done that it will put off your readers; worse still it might have the effect of giving the impression that your firm is a bit odd (and by that reckoning so is your work!).

The cost of such an advertising space – 6.5 x 4.5 cm for a local shopping guide – would be something in the region of £20–£40. One of their advantages over a newspaper is that they tend to sit around in people's homes, get picked up and generally have a longish life. Very often the publishers will offer to do the art work for you for a small charge, or even, if you can negotiate it as a new customer, for free!

Local advertisers

These are free and delivered to the door. The back pages are usually full of adverts for local builders, plumbers, joiners, etc. They have a number of plus points as far as your promotion is concerned.

- They get into a large number of homes.
- People will expect to find adverts for a range of building, joinery, plumbing and construction work – there are often at least three pages with these adverts.
- You can often do some good deals with the publishers since they are usually linked to your local newspapers.

If you phone the publishers up they will be often be keen to sell you a package. As far as costs are concerned, a 3 cm by 3 cm box advert will normally set you back some £40. e.g.

```
HOT RAY PLUMBING
    & HEATING
Free estimates. No call-
out charges. All work
guaranteed. All aspects
of plumbing/heating
      covered
  565-666 or 565 777
 mobile 0402 666 555
```

Yes it's a small advert and you might think you're not getting very much for your money. The trick is to try and buy into one of the packages these publishers often offer. For example, in the above, if HOT RAY were to purchase four of these 3 x 3 cm box adverts a month (i.e. one per week) the total cost would be £252 plus VAT (1998 prices). However, by putting one advert in the local newspaper on a particular day they nominate (usually a Wednesday), plus four weekly ads in the advertiser, the cost would be reduced to £168 plus VAT.

This is just one example; obviously these prices would change depending on the circulation of your local newspapers and the time of year. For instance it could be cheaper in July/Aug when so many people are on holiday and advertising dwindles compared with September/October when prices for ads for home improvement services tend to rise sharply as the nights draw in and people have longer evenings to plan all those wonderful changes to their homes, gardens, driveways, etc.

The important thing, referring back to our section on negotiation pages 120–23, is to find out what packages are available and negotiate. Remember, if you don't ask you won't get!

Thomson Local Guides and Yellow Pages

You can buy a space in one of the local guides or yellow pages. The prices will range from £50 to several hundred pounds depending on the size of spread you are looking for. You are less likely to be able to negotiate a package with these publishers but remember that their big advantage is coverage. They will be going into hundreds of thousands of homes and businesses throughout your telephone region which could contain several million people.

The Internet

Because of the rapid spread of the Internet or World Wide Web, firms are well advised to have a home page to illustrate their 'wares' and attract potential clients. There are plenty of consultants who will design pages for you – at a price. You can get a pretty good idea of what is required by browsing yourself through lists of construction companies. See what attracts you, adopt and adapt, but don't copy. Whatever you decide remember as with so much graphical communication:

- be clear (don't have the pages too cluttered or fancy)
- be accurate (don't make claims you cannot sustain)
- be visual (it is a visual medium so avoid masses of text).

Forms of sponsorship

There are hundreds of different ways of getting your firm's name in front of the public. First find your public. If you think that golfers are your public – i.e. they are likely to have porches built or conservatories enlarged – then spend some money on signs round your local course. The same advice would hold true for swimmers (local pool); tennis players (local courts); footballers (local ground), etc. Find your public and then attract them, e.g.

> Hole 3   335 yards   Par 4
>
> WALLS of Elborough for all your building needs.

Just remember that such a sign might hardly be noticed by a happy golfer. But if he or she had just smashed the ball from this sponsored tee into the woods or deep rough then your company might be on the receiving end of some expressive comments – balls to bloody Walls!

You may think that the passing motorist will be a better catch. Your firm might sponsor a roundabout (with flowers and well cut grass) so that all these admiring drivers can see what a wonderful outfit you are. Alternatively you might want to influence parents of your local school and offer to sponsor a prize for community service.

Local radio

Most towns and cities in the UK now have some kind of local commercial radio service. Rates for adverts vary but it is a form of promotion that you should certainly think carefully about. Although expensive compared with many other forms of advertising there's no doubt that it will spread your name over a large area. Remember, however, that when people hear an advert on radio they mostly don't have pencil and paper handy to jot down the details. This means that you will need to repeat the promotion using other forms – newspaper ads for instance – so that people will have your name reinforced in their memories. With local radio (and TV) and sponsorship, this is where we start talking about advertising campaigns which are outside the scope of this book. If you're interested then you should consult a public relations/advertising specialist. These companies can get good deals for your firm and can provide advice on overall strategy for your promotion which could include sponsorship – everything from a local football team to a pub darts competition! (You'll find these specialists in your local advertiser or yellow pages!)

Know what it is you're selling/offering and know it well.
This means doing some homework and, if in doubt, checking with colleagues. There's nothing less likely to put your customer off than a vague, half certain communication, on the lines of:

> I think...well it's supposed to . . .
> I'm not that certain to be honest . . .
> It's something like . . .
> We've changed things round recently and I'm not that sure . . .
> I may be wrong but . . .
> We've always done it . . . but now I'm not that sure . . .

Such vague phrases inspire little confidence. How would you feel if you were parting with good money for that holiday you've been saving up for so long and the holiday representative said:

> Well I think that's booked . . . you can never tell with this system . . .
> I could be wrong of course but you should land . . .

You'd be thinking of pulling out and placing your order with another operator.

You need to convince the customer that you know what you're talking about. The golden rule is: if in doubt about anything, then ask. Check with someone who does know or tell the customer that you will come back when you have more definite information. Never try to palm the customer off with vague promises or incomplete information. If you do you'll probably lose that customer and, as we noted earlier, you'll probably lose a few more through word of mouth reputation.

There is a key law of Customer Care! If you do give the customer a date for a delivery then 'Under Promise and Over Achieve', and never 'Over Promise and Under Achieve'. If you say:

No problem we'll get that order to you on Friday before 12 pm.

Then you really must mean this and not have in the back of your mind,

Well it'll be a real miracle if it gets there before next month.

It's so much better to play safe and quote Monday pm even though you are pretty certain that it will be delivered latest Friday pm. Then if all goes to plan you can phone up the customer and say, *'We've got that part in stock. Can we deliver it this afternoon?'* Your customers will be impressed with this quality of service. They will not be impressed if you over promise and if your promises seldom amount to very much.

What a shower! He told us Friday. Now we have to wait a whole flaming week before they can get it to us. Some service! I won't be going back there in a hurry.

Know the procedures.
Learn as quickly as possible how things run. Make sure that when a customer calls in person or on the phone that you understand what has to be done by you representing the firm. The procedures when taking the order may include:

- what do I have to fill in?
- what do they have to fill in?
- who do I pass this on to?
- what do I say to the customer about prices?
- what do I say if he/she requests a visit? etc.

Very often all you'll need to do is to take the customer's name and someone else will do the work for you, but there will come a time when because of absence and holidays you will need to understand the relevant procedures.

Follow up.

There's no point going to all this trouble, building up your records and making sure that you understand the procedures if at the end of the day you fail to follow up your investment of time and energy. If you get an enquiry and don't follow it up then not only will you lose that custom but your reputation may also suffer. You can imagine what your customer may think:

> Well I called and asked what they could do but they never came back to me so they can't be that interested. I'm not bothering any more with that lot. There's plenty of other plumbers in the town.

It is so surprising how many organisations never bother to follow up their leads. Sometimes this is due to carelessness, that odd note of a customer's name sellotaped to the end of the desk which inevitably comes off and gets swept into the basket, that hurried scribbling down of a message from Gary Scott which reads Harry Cott, that phone number 226-4747 which could be 226-4741, that address which looks like 5 Burway Ave but might be 6 Burnway Ave. It's all careless. Sometimes this carelessness it is due to pressure of work. In a small firm with callers and plenty of hustle and bustle, vital things may get forgotten. That key follow up call just never gets made, that letter is never replied to, that e-mail goes unanswered.

Sometimes it is because trade is good, business is flowing in, the phone never stops ringing and you say to yourself, 'Well *what's the point of following up these leads?'* Be prepared for bad weather. Don't just sit in the sunshine. Just because May, June, July and August were good months with plenty of work coming your way doesn't mean to say that the autumn or winter will be that good. When the business is flowing in, this is the very time that a successful business will realise that the good times will not last for ever and a recession may be on the way.

Very often carelessness is due to the fact that there is no real system for dealing with, logging and actioning enquiries. One way of making sure that follow-ups don't get lost or forgotten is to organise yourself a large year planner and put it up on the wall. Make use of brightly coloured flags or pins so that you can quickly see who you need to contact by which date. e.g.

Jan	Feb	Mar	April	May	June	July	Aug	Sept	Oct	Nov	Dec
Holiday										♦	
			♦		*Maint'ce*		*Holiday*				
	♦							♦			
											Holiday

- Remember to block off days/weeks when you will not be in a position to complete any work because of holidays or a shut down for maintenance/work on the premises, etc.
- But don't neglect it. If it ends up a mass of scribbled notes then the really important information will get buried and inevitably ignored.

These days there is all manner of computer software to do the job of your large noticeboard. The computer will send you signals – bleeps, flashing signs on your screen, etc. – to remind you to act on that promised call, visit or invitation to the premises. However, like any computer system it is only as good and as reliable as the humans who run, store and maintain the system.

If you do have a slack time with no work coming in then spend a day or two going through all your files and reading through all your customer lists. See which ones could be given a nudge – a gentle reminder that you are still in business.

Selling and negotiating skills with your customers

Selling – doing your SWOT

Having enticed customers through your efforts at advertising and promotion it is vital that when you actually meet them you manage to sell the product and negotiate the best possible deal. Selling is very much to do with communicating successfully with people – your prospects. We have already covered a number of the key points to do with communication skills on pages 17–25. Let us now build on these by focussing on some key selling skills.

It's quite a good idea when selling anything to consider SWOT. We looked at this on page 39. Remember it stands for

- Strengths
- Weaknesses
- Opportunities
- Threats or Barriers (then it's SWOB!)

This can be applied to an individual as well as to a firm. Let's have a look at a personal SWOT. This will be of great help when you prepare for interview for a new job (see page 38).

Activity

Take a moment now and jot down on a piece of paper what it is you think your key selling points are, i.e. the best cards that you can play.

Here is a list to provide you with a guide.

Strengths
- Skills that you've got and at what level.
- Qualifications – NVQs, SVQs , City & Guilds, other certificates.
- Experience you've had with different jobs. What kinds of experience have they been? Have they been much the same in different jobs or have you had a wide range of different kinds of experience?
- Do you have any recommendations from previous customers?
- What about your reliability? Do you have a good health record – e.g. you've never had to cancel/postpone a job?

Weaknesses
- I've only been in business just over a year.
- There's only one other apart from me and he's only been in the trade six months.
- Most of the jobs we've got have been local and on the small side.

Opportunities
That course at college I'm doing should provide me with a chance to find new business. There's some good contacts in the class I'm in.

Threats/barriers
There's a lot of other small construction businesses in the area all looking for work.

This kind of analysis can be a little disheartening, especially when the threats/barriers list is that much longer than the opportunities and strengths one! However, we certainly recommend that you try it. Most people never seriously sit down and try to estimate their own strengths; they are far too busy worrying about their weakness or the various threats that may hit them. At least going through a SWOT exercise will give you a very good start when it comes to setting out a case to the bank manager when you wish to go for a business loan.

When you have established your personal strengths you then have to work out those of your firm/business and sell them to others. The SWOT for the firm follows very much the same lines as the personal one. Here is one for a small family run builders.

Strengths
- Has been established over 20 years.
- Has regular repeat business.

Weaknesses
Relies on father and son, father due to retire next year. It will be difficult to find a replacement. When father goes so, the fear is, will some of the older customers.

Opportunities
Your father's retirement could be the chance to develop some new markets which he's always been against.

Threats/barriers
- Local competition.
- Few new homes being built in area.

- Rising unemployment.
- People cutting back on spending.

You can see how this SWOT analysis can be useful when it comes to future planning. Any firm that doesn't do some kind of analysis like this runs the risk of being left behind.

Selling technique

Here are some key points of sales technique. We have already established the importance of knowing your customer, knowing your product, etc. We have also looked at the essentials of inter-personal communication – establishing and maintaining eye contact (page 17), active listening skills (page 35), etc.

Activity

Jot down what are for you the most important factors that make a good sales technique. Think of shops you go to, particular salesmen and women. List your top five.

With your top five in mind see if you agree with our list.

Being friendly, helpful, organised and clean in appearance
You obviously want to make a good impression on your potential customer. Remember the various points covered on pages 35–7 where

we examined, amongst other factors, eye contact and tone of voice. You'll certainly need to:

- look at your customer and not be busy with machinery as you speak;
- sound interested not bored and fed up;
- shake hands and not put off the customer with oil/grease, smell, etc.

Wearing a reasonably fresh overall without too many stains, rips, patches and broken zips will help create a more positive image. If it's too clean then the customer may think that you're not doing any real work!

The ability to listen to the customer
All customers think they have special needs. Your job when selling is to listen carefully. Even though you've heard the same tale before, you still have to listen. You cannot sell if you don't listen for:

- their names. Use names when you reply to questions but be careful, especially with older customers, that you don't go straight into first names as in, *'Well John I think we could do a deal here.'* You'd be better saying, *'Well Mr Jones . . .'*
- the exact nature of their needs. You'll find it useful to repeat this information. As we said about active listening (page 35) summarising/repeating key information back to the other party indicates that you have been paying attention.

The ability of the salesperson to put himself/herself in the buyer's position
This is the ability to see the service or product from the buyer's point of view. When selling, ask yourself:

- are the benefits of my product clear enough to my customer?
- have I presented the advantages of mine over others well enough?
- have I shown how the product/service fits in with their 'special' needs?

All buyers think they are special and that they have special needs, so your potential buyer wants to have these questions firmly answered; you've got to put your goods in the window and not keep them on the back shelf!

The ability to counter the buyer's possible objections
If you don't then he or she is much less likely to buy. You need to be able to address any weaknesses/shortcomings in the product or

service. You know yourself that when you are buying a motor bike, car, CD player, etc. if the salesperson frankly describes any drawbacks and yet emphasises the strengths of the product you will probably feel more reassured, compared with the situation where nothing is said about any weakness – all is praise and perfection – and yet you know that it can't be all good news. There must be some less than perfect features, some drawbacks. You may think to yourself: *'Why is he keeping so quiet about this?'*

We certainly don't mean that you deliberately talk down your product. But you should appreciate that most customers are sensible enough to realise that there are advantages and disadvantages in any purchase – your job is to get them to see that any disadvantages are easily outweighed by the very obvious advantages.

Prepare to talk about any weaknesses (and try to turn them into strengths) without entirely selling yourself down the river. e.g.

Yes, this is more expensive, <u>but</u> the difference between this and the cheaper one is . . .

You may have to wait longer for this particular one to dry <u>but</u> you'll find that it . . .

It's heavier than this one which may be a problem <u>but</u> there's no doubt it'll last much longer than the other will . . .

We sometimes have to renew these joints <u>but</u> that's a much easier job than with these ones where we have to take out the whole frame . . .

If you see your potential buyer looking/sounding uneasy about something you're saying then you have to try and reassure them. There's no point in leaving gaps, unspoken worries and doubts about you, your firm or its products/services.

The ability to provide rapid, easy to understand information on prices and terms of sale
Nothing stops a sale in its tracks more than if the potential buyer can't get a quick and easy to understand idea of the price he's got to find. How many times have you been puzzled when the salesperson reels off finance terms, APR, easy to pay deals with part finance, etc.? All the time you're desperate to have a nice, clear, plain English explanation and find out what it's going to cost.

The lesson is keep it simple, don't bore the customer: and above all know what you're talking about. Remember if you are in any doubt about the financial aspects of the sale, stop and ask or tell the customer you'll check and come back as soon as possible, and be sure to do that!

This is the point in the sale where you can negotiate (see pages 120–23). This means that you have to understand fully the various prices and deals on offer. This probably means that you have to ask and check with your colleagues. It is fatal for your sales prospects if one of your colleagues quotes one set of prices and a few hours later the potential customer gets another set from a different colleague. This will give a very poor impression of you and your firm.

The ability to follow up from the first contact
We saw how this was important in customer care. It's vital in sales. The aim is to hook the buyer so that he or she will stay with you not just for the one sale but for many. If you don't follow up the prospects then they'll swim away to another firm. Hook them by that follow up phone call or letter. It will show that you are serious about the sale. e.g.

> Is that Bob Smith? This is Harry Jones from Halls the Plumbers.
> I said I'd get back to you with further details on that job you were thinking of doing. I've done the sums and we can now offer you . . .

Getting back quickly to the customer – and by quickly we mean that same day or first thing the day after the meeting – can create a favourable impression. Learn how to turn round your paper work – estimates, etc. – rapidly. It gives the impression not only that you are keen to get the business but that you have efficient systems. Efficient turning round of the paper work means that your potential customer may believe that your construction work will also be efficient.

Remember that even if your sales efforts do not result in immediate business, if you have presented yourself and your firm well and made a positive impact on the other person he or she might well remember you and come back to you at a later date or even recommend you to a friend.

The ability to deliver on the deal
Where's the beef? There's no point in having all talk and no action. The customer will want to see results, improvements, the service starting, the plans being drawn, the pipes repaired, etc.

Negotiation

We've already mentioned the need to negotiate when it comes to prices. We need to run through some of the key ideas and skills in negotiation. But first try this.

> **Activity**
>
> Think back to any situation where you have tried to do a deal with someone.
> What technique/s did you use?
> Did you get the deal you wanted?
> If not why not?

Here then are some of the key techniques but do remember that basically everything is negotiable except death and taxes, and even with the Inland Revenue you might be able to negotiate exactly when you pay up (they'll charge you interest of course for the privilege!).

Clear aims
You need to set yourself some clear aims; i.e. what it is that you want out of this negotiation. There's no point trying to negotiate anything if you don't have these aims firmly in your mind. You will soon be blown off track. For instance do you want to gain:

- a higher price
- a refund
- a delay
- a new contract
- a change in terms
- fewer payments, etc.

Bottom line
You must have a bottom line. This means that point in any negotiation where you say 'No thanks. No deal.'

Suppose you want to sell your car. You advertise in the evening newspaper classifieds section, and ask for £1000 o.n.o. (yes it's a great car!). Someone comes along and you start negotiating. Your aim is obvious – you want to sell the car. But what is your bottom line price? O.n.o.(or nearest offer) could mean anything. Would you, for instance take £500 or even £450? Presumably when you set the price at £1000 you must have had a figure of say £900 as acceptable. In other words £100 was your negotiation range. Would you take £800 or would you say no at anything below £850?

Can you see the importance of having a bottom line? If you have this clearly set then you will be in a much stronger situation when it comes to all kinds of negotiation. You will be much less likely to be blown off course, get a rotten deal and then feel miserable about yourself. Why did I take that rotten price? The car was worth far more than that!

Your bottom line has to be realistic. It's no good going for £950 when you know that any reasonable offer will be in the region of £750–£850. Having a realistic bottom line means that you have to do some homework, some preparation before you start the actual negotiation. Returning to your car, if you knew that you could get £100 for spare parts that might influence your bottom line calculations. If you also knew that for £275 at a local garage you could improve the performance of your car so much so that it would be worth your while keeping it for another year that might affect your negotiation. With this kind of information you would be in a better position to refuse any low offers. Your bottom line would be more secure and therefore your confidence increased.

Win–win
Negotiation is all about arriving at a win–win situation – you win and they win; that is the other party gets something out of the deal. Negotiation involves some measure of compromise. You will seldom, if ever, get all you want. If you do it is likely that the other party won't get anything of what they're looking for. If this happens then it is highly unlikely that they will want to negotiate with you or your firm again. It's the difference between aggressive and assertive behaviours that we noted on pages 54–6.

We've already talked about customer care. One way to keep your customers is to negotiate on a win–win basis, i.e. both sides get something out of the deal. If you grind a customer down with a really tough negotiation – a win–lose one – then they are very unlikely ever to be your customer again. Think how bitter, angry, generally fed up you've felt when you've come out of a deal feeling you were stitched up, got at and given the run-around. We're sure that you vowed never to do business with that party again; well those will be the feelings of your would-be customers!

Creative thinking
What is needed in negotiation, apart from preparation and a bottom line, is some creative thinking. There are bound to be situations where both parties in the negotiation are stuck. There is no likelihood of any deal emerging. This is where you have to make proposals. For example, going back to your car deal, if both parties are stuck (they're only going to £700 as their absolute maximum and you're not budging from £800 as your minimum) then how could you arrive at some kind of win–win position? You might:

- make him an offer to accept payment in two instalments, say £400 this month and £350 the next, the car being handed over on the final payment;
- suggest dropping your offer to £825 in return for cash up front.

These may not seem like particularly attractive or practical suggestions but they offer some way out of the deadlock, some way of breaking the log jam.

Keeping a record
Finally, in any negotiation you need to be quite sure at the end what has been agreed. All too often negotiations fail because both parties come away with different conclusions. This sort of thing is all too common.

> *I thought you were going to pay me £400 this month and £400 next month.*
> No, I said £500 this month and £400 next. Surely that was clear?
> *It wasn't to me!*
> Well it should have been. We talked it through long enough.
> *That's what you remember. I distinctly heard you say £400 for both instalments – it makes sense after all.*
> Not to me it doesn't!
> *Well take your car – you know what you can do with it.*
> That's the last time I try and do a deal with you.

The lesson from this is clear: memories of what was said during a negotiation are very slippery. Write down what it is you've agreed and make sure that both sides have a copy of the agreement. Make sure that these are signed by both parties.

Sending out estimates and invoices

Estimates

Sending out an estimate is a very important aspect of your communication with your customers. An estimate is precisely that – an estimate. You must make it very clear to your customer that it is an estimate. That word needs to be printed in bold at the top of the page. Here is an example of an estimate.

JOB ESTIMATE

Cast Iron Construction
14 Bolt Road
ELBOROUGH
EB 13 4RE

Phone No 0112-334-5656
Fax No 0112-334-0989
Mobile 041-667-7895

To: Mr T Jones
24 Newmills Rd
Elborough

Job Description

To replace old lead Hot & Cold water supply
pipework with new copper pipework, supplies
insulated and fixed to brick walls.

ITEMISED ESTIMATE: TIME & MATERIALS	AMOUNT
Materials cost	£988.08
+ VAT @ 17.5%	£172.91
Labour cost	£426.15
+VAT @ 17.5%	£74.58
TOTAL ESTIMATED JOB COST	£1661.72

This is an estimate only. This estimate is for
completing the job as described above, based on
our evaluation. It does not include unforeseen
price increases or additional labour and
materials which may be required should
problems arise.

Prepared by *Jim Bolt*

Date Jan 13th 1999

It is essential that you include a phrase such as the one above. It should set out the limits of the estimate. You do not want an angry customer complaining that the final bill has no relationship to the estimate you submitted. Err on the side of caution. Remember: never over promise and under perform, better to under promise and over perform.

Activity

There is one omission from this estimate. Can you spot it? Read it carefully again.

Somewhere, usually at the bottom, should be set out the time limits for such an estimate. e.g:

- this estimate will operate for three months from the above date, or,
- this estimate will hold good for three months from the above date, or
- this estimate is valid for three months from the above date.

Three months is a reasonable time to hold an estimate. You can never predict when your costs are going to rise. Three months is also a gentle reminder to the customer not to hang about. You want him or her to commit to the expense. You want to be able to get the work entered into your calendar of work for the year.

You've got to think about your cash flow. You don't want to be borrowing from the bank to pay your fixed costs (the ones you can't change such as council tax, wages, etc.) if you can help it. You certainly don't want a whole lot of estimated jobs all coming together in one period and then a long period of slack time. You need to be able to spread out the work. So use the estimate as a way of doing this.

Invoices

No one particularly likes receiving an invoice – a bill. Your customers won't be that pleased to see your name and the amount of money they owe you. However it is all part of communicating with your customer and staying in business. Here is an example.

```
                    W. A. MOPUP & SONS
           Slaters, Plasterers and Building Contractors

              21 Railway Cuttings. Elborough EB23 4GH
              Telephone 01201-223-4567  Fax 01201-223-8767

   To: Mrs Jones
   15 Clark Road
   ELBOROUGH

   VAT Reg No 356 0098 56            Tax Point 11.8.98


   Carry out temporary roof repair

     Labour charge                               £185.00
     6m² Flexshield Plain Felt @ £4.00 per m²  £24.00
     Gas                                          £1.60
     4 Slates                 @ £1.00            £4.00
     Cement sand                                  £1.50
     Transport charge                            £10.00


                                                £226.10
     VAT @ 17.5                                   £39.56

     Total                                       £265.66

   Settlement of this invoice is due by Sept 12th 1998
```

Do make sure that you check all figures and your addition. Remember to check whether or not to add VAT depending on your business turnover.

Try and make the invoice as easy to read as possible. Line up the various figures so that they can easily be added up. Avoid such words as miscellaneous or extras. They make customers suspicious. Try to spell out what they contain. How would you like it if at the end of the garage bill for servicing your car they added miscellaneous or extras and slapped another £8.50 on the bill. You'd like to know what these words mean. Keep it simple and clear. Remember you may be the one who has to handle the complaint from the angry and bewildered customers who can't understand what's in their invoice.

Since November 1st 1998 small businesses have had a right under law to add interest on their invoices to those customers who pay late. However before this can happen you as the supplier of goods and services have to do the following:

- Make sure that the invoice is clearly dated.
- Ensure that your settlement terms are printed clearly on the invoice (i.e if you say *within 30 days*, does that mean 30 working days? Since Nov 1st 1998 you can add *interest will be charged on late payment*. At the time of writing this is bank rate + 8%.
- Ensure that it is sent first class post.

Handling your customers' questions, complaints, etc.

Much of your communication with your customer will be handling questions and complaints and dealing with awkward situations and people! Let us examine these in turn.

Questions

These can reveal several things.

Lack of clarity

Your customer may genuinely not understand something. This may be to do with the price, an aspect of service, materials to be used, guarantees offered, etc. If you get such questions, particularly from more than one customer, then it is likely that there is something about the way that you explain things that you will need to alter. Perhaps you are not clear enough and this lack of clarity muddles up your listener. It may be that you talk too much, provide too much detail and that you would be better to write things down for the customer to take away and refer to later.

Concern about your firm

The questions may show real concerns which your customer does not feel able to express openly, e.g.

- Has most of your firm's work been with modern housing? This question may reveal that your potential customer is a little concerned that you may only know how to repair modern roofs and haven't had enough experience of pre-war ones.
- Has all your firm's work been local? This innocent sounding question may reveal a serious worry from the potential customer that you don't appear to have landed any contract outside the neighbourhood, i.e. that you can't be taken that seriously in regional or national terms. It may also reveal a doubt as to whether your firm could take on a bigger job than usual.

In the section on active listening (pages 35–6) we saw the need for careful listening to detect any underlying meaning expressed through the *way* things are said. Listen hard to your potential customer. Watch him or her very carefully. See if you can detect these worries and do your best to patch up any weaknesses.

Activity

How would you have responded to these questions?
Read them again before you have a look at our replies.

1. Has most of your firm's work been on modern housing?
2. Has all your firm's work been local?

Question 1

There's little point in lying if most of the work *has* been with modern housing. What you need to point out to the customer, who may not feel that you have the skills and experience to work on their 1920s house, is that modern housing comes in all sorts and sizes and the demands on your skills and technique have been considerable. You may not convince the customer but at least you can try. Show them pictures of the houses you've built, the workmanship, etc.

Question 2

This question about where your work comes from indicates a worry that you've never made a regional or even national level. Talk about the reputation you've built up, the amount of repeat business and how you've specialised in work from this particular area – knowing your customers has been more important to you than spending a lot of money advertising round the country. Again it might not be convincing enough but at least you've made the effort.

Handling complaints

All firms get complaints. It's part and parcel of business life. People are encouraged now more than ever to complain and not to take bad workmanship, etc. lying down. The various professional institutes – for plumbing, building, etc. – have their codes of practice which registered firms must follow when it comes to handling complaints. Some complaints can be tackled within minutes; others may require the services of a lawyer and take months to resolve.

The important question is how well do you handle complaints

when they come, as come they will? It is a well known fact of business life that those firms who handle complaints well not only improve their reputation with existing customers but may well attract new ones on the strength of this. You know the sort of thing. One customer says to another:

Well they did make a mistake but I must say I was very impressed *when* they put it right within a couple of days and no fuss or bother. I got my money back straight away. Compared with most other places that was good.

Would customers who have had complaints say this about your firm? Or would they grumble at the slow response and the lack of concern shown?

Activity

Think about a complaint that you have made to a company/organisation/local authority/tax department, etc.
Was it well handled i.e. what did they do to reassure you?
From your experiences what would you suggest are the five most important things to remember when it comes to the efficient handling of complaints?

Here are some key aspects to compare with those you've selected.

Handle the complaint speedily.
Don't hang about and leave the letter in a file. Answer it as quickly as you can. If you don't have all the facts then send a holding letter. Often the longer the complainant has to wait the angrier they become. You can help diffuse their anger with a speedy response.

Activity

What do you think of this letter in reply to a customer's complaint? Imagine that you were the person who had made the complaint. Would you be happy with this?

```
          UP UP AND AWAY ROOFING REPAIRS
             42 Nailstick Rd Elborough

Mrs M Smith
15 Albany Terrace
Elborough                          June 25th 1999
EB3 6AT


Dear Mrs Smith

             Re: Work on Your Premises

    This is in reply to your letter of June 23rd
  1999 in which you complained about the mess
  left behind by our firm after work on your roof.
  Those who were involved carrying out the
  repairs are working away from the office in
  Northants at present. They are due back here
  next Monday.
    I shall be meeting with them on that date and
  hope to be able to clarify this matter.
    I shall get back to you by the end of next
  week.

                  Yours sincerely

                  Jim D Smith

                  Site Manager
```

In many cases a telephone call will be enough but many people are ex-directory or simply out when you call so that a letter or fax is often better. Increasingly more and more people will have e-mail so use it. A communication such as the one above allows you to buy some time to check the facts, talk to those involved and generally prepare your case. The customer should be happy provided that you do get back within the stated time.

As with successful negotiation, so successful handling of complaints often boils down to preparation – knowing your ground, finding out what actually happened (as opposed to what was supposed to have occurred), talking to the people involved, establishing the key facts, etc. Never admit blame or fault in a situation like the above without first checking on the facts from your side.

Communicate the issues clearly.
Complaints are often badly handled because the basic facts are never clarified – the *who* is supposed to have done *what* to *whom, how, where* and *when!* So get these essentials clear. This is where a customer complaint form is useful. It sets out a consistent way of recording the essential information. Here's an example.

Customer Complaint Form

Date of Complaint   June 23rd 1999

Parties involved   Mrs M Smith of 15 Albany St Elborough

Nature of complaint   Mess left behind after working on her roof, specifically debris, tiles, nails, etc. She apparently had asked for this to be removed on last day of working.

Action taken   Holding letter to Mrs Smith. Talked to workers involved including foreman. Admitted there had been rather a rush on the last day and they had perhaps not cleared up as well as they might have. Sent squad on Tuesday June 29th pm to clear up. Phoned Mrs Smith and apologised for this.

This kind of form can provide important information. It acts as a record of what took place. OK this is only a fairly minor problem and one quickly put right. But imagine if it had been to do with leaving asbestos around or some other dangerous materials? Think of the possible legal implications for you and the firm you work for! If it ever came to a legal action it would be important to have good records of what happened.

More and more people are going to court if they are not satisfied. You need to be prepared for this. Keep records. Keep copies of all letters and invoices. Keep short notes on calls made. Keep them in a file not on scraps of paper that will get lost.

Activity

Reading though the above form what is missing from it?
Have another careful read.

You might have noticed that there is no space on this form for conclusions/lessons learned. The so what as opposed simply to the what. This leads us to the third essential when handling complaints.

Learn from the experience. Don't repeat the same mistake.
What's the point in going all through this fuss if there are no lessons to be learned from it at the end? The lesson here is that time just has to be set aside on the last day for clearing up to take place. Think of the expense of having to send out a team to clear up Mrs Smith's premises – the extra journeys, etc. when it could all have been done first time round. Think of the negative publicity aspects. Mrs Smith is sure to tell her neighbours. Some of them will remember her words and some of them will pass this on to their friends and neighbours!

Do your very best not to lose your cool!
Complaints are seldom pleasant to receive. The complainant is often angry, upset and hot and bothered. This means that it is essential that you, on the receiving end, keep cool and don't allow your anger to show. Keep your feelings in the Adult – don't let the Critical Parent take over! If you do boil over then there's very little hope of resolving the complaint. Try and keep cool on the outside even though you are bubbling with anger on the inside. Think of the problem and not the person. It's very difficult but essential if you are to avoid blowing up! Here are some tips:

- count ten under your breath to stay cool;
- concentrate on the issues not the person;
- summarise and check back with the person on the essential facts:

A fracture in one of the seals, and it was Saturday you noticed it.

- play for time; don't give your verdict until you have checked all the facts:

I'll have to check back on the records. I'll phone you tomorrow first thing.

Do keep your promise and get back to that customer.

Communicating with your customer on site and in their homes

On site

Remember that you are more or less at home on the site. Depending on how much time you've spent there, you will know your way around the place. Your customer, as a visitor, will not. You may have become accustomed to the noises, the vibrations, the general clutter and disturbance but it will all be very new to your visitor. This is why we have to be very careful about the health and safety of any visitor on our sites. There are several issues here.

- Firstly, visitors on site should follow the warning and advice notices which must be put up for them and which should state in very large bold print:

> **ALL VISITORS REPORT TO SITE OFFICE**

- Secondly, visitors should be accompanied on site and never left to wander about on their own. They should be given a visitor's badge and a hard hat. This business of accompanying visitors around the site is often very difficult to achieve but it is essential for everyone's safety. If you see people who are looking lost then approach them and ask if you can help them find their way.

Remember that thieves coming to the site on the off chance, or deliberately spying out the land, may walk about with confidence, so be prepared to 'challenge' anyone who you do not know. One of the best challenges is:

Can I help you? Are you looking for anyone in particular?

If they can't provide you with a name then warn the site foreman – it's always better to be safe than sorry when it comes to site security. You can have all the latest CCTV equipment but the best form of security is for all workers to keep their eye open so that anything unusual will be noticed.

- Thirdly, if you have to talk to a visitor about some aspect of a job then it is often much better for both parties if this can be done inside a hut, site office, etc. You may be able to hold a conversation in the open air with all the noise and bustle, he or she may not be so comfortable. Besides there may be papers to read and items to sign. For this you'll need some kind of desk. Remember Customer Care!

In their own homes

Activity

Have you ever had work done in your house?
Think back to your feelings when the 'gang' arrived.
What would you like them to have done to make the experience more pleasant?
List these things.

We suggest that for most people these are the concerns they have when workers start in, around, under or over their homes.

- *Loss of privacy.* Just having other people about when you're at home can give rise to a sense of 'being invaded'. These feelings used to be the preserve of 'house-wives' and retired people. Today more and more people work from home and of course there are more house-husbands.
- *Worries about security.* Having other people about their house will make most people, especially those who live on their own, a trifle insecure.
- *Noise.* The sound of hammers, drills, saws, cement mixers, etc. can be very upsetting if one is not used to them. The other aspect to noise is the sound of radios blaring, very often producing the sort of music which would make many customers want to curl up with pain. Try and work out some reasonable compromise with the customer in terms of the music's loudness and duration, i.e. how long it will be on for.
- *Mess, dust and smells.* These are bound to be created to some degree as work goes on in and around the house.
- *Dangers to occupants* from slipping on temporary floors, stairways, etc.

Activity

What can be done by those in construction to minimise these disturbances? Have a think and jot down up to five points.

Let's divide this up into before work commences, during the actual work and after the work has been completed.

Before
This is where we need to communicate with our customer. If things are explained beforehand as regards possible disturbance, then some of the sting will be taken away. It's when these 'invasions' suddenly arrive that the most pain is caused. We need to avoid these reactions:

Why wasn't I told about all this dust?
Why didn't someone from the firm tell us that we'd have to have all the windows open?
Why didn't someone come and talk about car parking instead of my waking up one morning to find two cars and a van on the front lawn!

Most people are reasonable – most people! If the reasons for any disturbance are explained to them then most will understand, especially if it's only for a short time.

If they are involved in the planning for what is going to happen they may well be able to contribute and minimise the difficulties. They will come up with suggestions for parking, opening back doors, making available downstairs toilets, providing some dust sheets, protecting special bits of furniture or paintings, showing you where the water and gas mains are, etc.

It will be so much easier to work on site if you can gain that precious co-operation with and trust from the customer. It may appear to some in the world of construction to be a waste of valuable time going round to the customer's house before work starts. There's all that phoning, trying to get in, having to make a special journey, etc. We understand this but we also realise that if you can spare just a few minutes for face-to-face talking to the customer and their neighbours before work actually begins you are likely to save a great deal of unnecessary fuss and bother. You are also helping your firm's reputation. This kind of customer care does so much to lift your reputation. People will mention it to their friends; word will get around that you are a considerate and helpful firm. People will not mind the inevitable mess and noise if they know what's going on.

If you come along intruding into their privacy and into their own space without so much as an excuse me or I'm sorry about all this, then you will get some pretty browned off customers. Each browned off customer will tell another five. You can forget that money you spent on local advertising – it's being done for you but not in the way that you want!

Here is a checklist of what needs to be clarified with customers before work starts in their homes.

- Starting and finishing dates to fit in with customers' requirements, firms' needs and holiday arrangements, etc.
- Parking. Where do cars and vans park? Can use be made of a drive? Is there a neighbour who would be willing to let his/her driveway be used between 8.30 am and 5.00 pm on weekdays?
- Toilets. Is there a downstairs toilet that the workers can use?
- Provision of dust sheets, overlays on carpets?
- Particular periods of the day where if possible noise could be minimised, e.g. lunch breaks, getting young children to sleep for afternoon nap, etc.
- Rules on smoking in and around premises.
- Use of kitchen facilities and telephones.

- Use of radios – loudness of music being played, etc.
- A collection system for debris caused – placement of black bags, etc.

Activity

From your experience can you think of any more issues that should be made clear? List these.

During
Make sure that there is a named person who the householder can go to for advice, to offer a suggestion or to raise a complaint. The important thing is to keep the channels of communication open and to try and resolve small difficulties before they grow into big issues. Make sure that this person is known. For instance:

> Mrs Brown this is Harry Jones. Harry is your point of call for any problems and worries that you may have during the time the building work is going on. Let him know and he can do something about it. He'll be here most days. If he's not around then leave a note for him in this box – one of the others will get it to him as soon as possible.

Now that's better than a grunt to Mrs Brown as the gang go in with pick axes and sledge hammers!

After
- Check before finishing the job that all rubbish has been cleared away.
- Make sure that any damage to the premises – scraped paint work, chipped stonework, dug up lawns or flower beds – either has been repaired or some system is at hand to put them right at a later date.
- Record any customer complaints. These must be recorded (written down) and not just talked about. Take any complaints that can't be ironed out there and then back to base to be properly dealt with.

After sales service

There's little point in being a great salesperson and having a wonderful product to sell if after concluding the deal there's no follow up. By this we don't mean getting on the phone every day and badgering the client. Follow up means making sensible use of your customer record card (pages 100–101) or sensible use of customer surveys (page 105) – generally keeping in touch.

There's always a very fine line between just keeping in touch and badgering and annoying the client. Here are a few reasons which you might use to get back in contact with clients.

- Your records tell you that you promised to be in touch at the start of the year. It could be that some maintenance check will be required.
- There's an upgrade of some kind in the product or services you offer and you felt it would be sensible to inform your various clients.
- You are changing the nature of your service/product, i.e. moving out of one type of plumbing service and into another. This provides an excellent opportunity to get back in touch.
- You are moving your premises, changing your phone/fax numbers, opening an e-mail or WEB site and want your clients to know.
- You have won some award or completed some challenge. This is newsworthy and should be 'sold' to customers while the news is 'hot'.
- You are raising money for a charity, sponsoring some event and invite others to join you.
- You are offering special discounts – end of year or summer sale.

With large numbers of clients it might be well worth your while thinking about a newsletter which you can send out with your 'news'. Such a newsletter has the advantage of saving you the time it takes to contact each client individually. For a modest investment you can have something produced which is good looking and will create a favourable impression with the reader.

The main disadvantage is that there are so many newsletters, magazines, trade brochures and glossies coming through the post (and increasingly down the fax machine) that your effort on which you have spent so much time, money and effort may often be consigned (after a rapid glance if you're lucky) to the waste paper basket. You might be better to do some gentle calling and telephoning.

Remember that if you have left a good impression (and a business card) the customer will in all likelihood get back to you. That's what you really want.

Group exercise

In your groups consider how you as individuals have been treated as customers by any organisation. (It could be one from the construction industry.) Sort out those experiences where you feel:

(1) the communication between the organisation and you the customer was excellent;
(2) the communication was OK, reasonable;
(3) the communication was disappointing and poor.

You might like to consider in your discussions:

- the initial contact, i.e. the first phone call or letter;
- the speed or lack of it in handling your business;
- the general politeness of those involved (or lack of it);
- the quality of the paper work – readability of letters, notes, etc;
- the ways in which any complaints/suggestions were handled.

Talk through your results with other groups and your tutor. Consider what pointers there are from this exercise for your work in the construction industry now and in the future.

6

Communicating with authorities and agencies

It's that ... tax office, I can't make them understand that I've already paid that money.

But I'm no longer the owner of the car.

Fill in which form? I've already filled in two.

Apart from all those individuals you will be required to communicate with – those on site and your customers – you will also from time to time have to communicate with outside agencies and authorities. By authorities and agencies we mean all those bodies that collect money from us – Inland Revenue and National Insurance; those that ensure that we follow the law – planning authorities, health and safety inspectors; and those who we need to communicate with in order to get vital pieces of paper such as licences for cars and lorries, etc.

Communication with such authorities and agencies can be awkward as those comments above suggest because:

- they know more than we do – they're the experts;
- they can sometimes appear to be off putting and this can makes us feel nervous;
- they are staffed by people who are often under tremendous pressure – they may sometimes react in what appears to us to be a rather cold and off hand way;
- they often use special terms, abbreviations, etc. which can be confusing for us, e.g. completion certificates; naturalised as a British national; registered keeper.
- we are often unsure as to their procedures, their way of handling things, this can make us feel uncomfortable;
- we often have to communicate with these authorities and agencies when we are under pressure to get things done.

Because of these factors we need to be well prepared in terms of our communication with them. What we say and what we write can be very important for us in getting our money, getting that licence, that passport and that planning application. These communications are not to be skimped. We need to take our time and get them right otherwise we face delays and upsets which usually means costs and loss of business.

Although we may not look forward to such communications we should remember that over the past 10 years these agencies and many authorities have made a very real effort to train their staff to be more polite and considerate. They have set themselves standards of performance. They are trying to put the principles of customer care into practice. So to help them and help yourself here are some hints for you.

Communicating on the telephone

Think back to those points we made on pages 40–42 when we looked at applying for a job on the phone – prepare by getting the essential facts handy. In particular:

Keep calm and be prepared.
Remember the busier the person at the desk in one of these agencies the more we can help by staying calm, cool and collected, by getting our facts clear and by providing the essential bits of information such as names, reference numbers, codes, addresses, etc.

It may be your first call to this agency for months but to that person on the other end of the line it could represent the 75th that morning! How would you feel faced with that kind of a barrage especially when not all of those incoming calls will have been made politely!

Activity

Read the following extract from a telephone call to a government agency.
What advice would you give to the caller as to his or her communication?

Hello I'm calling about my contributions. I've had this letter which
says I owe you £130 to make up what I should have paid.
Excuse me can I . . .
There's no way that I haven't paid. I've got all my records and . . .
Excuse me. Can I have your name.
Jones, I've always paid class 2 contributions on the dot. It's typical
of you the way you treat self employed people, you don't
appreciate . . .
I'm passing you on to someone who can help you. Just a moment.
Hello. I've just been telling your colleague.
Can I have your name and National Insurance number please?
W A Jones and its . . . YA 14 06 00 B.
Is this an enquiry into your contributions?
Too damn right it is.
*If you'll just wait one moment I'll look up the details on the screen.
Jones W A of 15 Thurock Rd, Elborough.*
Yes, that's right.
According to our records your last payment was . . .

Finally, we start really communicating but only after Mr Jones has
increased his blood pressure considerably and for little in terms of
result. He made the great mistake of directing his anger and frustra-
tion at the first person to answer the phone. It would be much better
to wait and get himself directed to the unit or sector that actually deals
with his case.

Compare this version.

Hello my name is Jones and I'm phoning to query a notice I received
from you which states that I'm behind in my contributions.
Hold on please. I'll transfer you to the section that deals with this.
OK (pause while he is re directed).
Hello. Can I have your name and national insurance number?
It's WA Jones and the number is YA 14 06 00 B.
*Just wait one moment while I get the information up on the screen.
According to our records you last paid on 15th March 1997.*
Well I make it that the last payment was 15th of June.
*I'll check on that. Have you a phone number I can get back to
you on?*

We're now getting somewhere, some real communication – both parties actually listening to each other rather than having a shouting match.

We mentioned on page 40 that it is important when telephoning to check that you've actually arrived at the right place. With agencies on the scale of the DHSS, Contributions, Tax, Health and Safety, etc there is absolutely no point in launching into your big speech until as we said you've navigated your way to the department, desk, sector, unit, sub-unit that has your file.

It is essential to give your name and any reference number. The Contributions Agency in Newcastle has over 20 million names in its files; it's probably got close to half a million Jones and several thousand W A Joneses. So always have your reference number to hand before you make that call. Just giving a name won't be enough.

Be assertive.

You need to be assertive on the phone; there's no point in being all meek and humble. Think back to our section on assertiveness on pages 54–6. Remember that see-saw balance. Try and stay in the assertive part and avoid moving to the passive or, in going all the way to the opposite end of the see-saw, the aggressive.

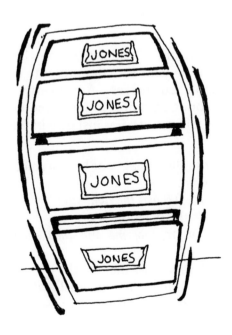

It is often when we are dealing with officials over the phone that our blood can boil. We often see them – those on the other end of the line – as bloody minded, obstinate, pig-headed, interfering, bound by rules and regulations, not able to make commonsense decisions. If we communicate this attitude in our manner – our tone of voice – then it's more likely that they will think that we are the nuisance, the trouble makers, out to be deliberately difficult. They will often react accordingly.

This telephone conversation is what we mean by the aggressive. It uses the sledgehammer method to crack a nut.

Hello. I'm phoning about my tax.
Yes. Can I ...
You've no damn right ...
Excuse me ...
... putting me to all this trouble.
Sorry can I ...
I'm writing to my MP about this.
If you could please let me ....
I can't afford to pay this amount. It's not fair. Why don't you go after the big boys who are cheating the system out of millions rather than bother the likes of me, trying to keep a small building firm going. How would you like it?
I'm sorry. Could you please give me your name and reference number.

This is aggressive behaviour. It will get you nowhere. The operator is not in a position to argue your case or enter into debate with you on the taxation of small businesses. An assertive rather than an aggressive style of communication to the same tax office would be much more likely to gain results. For instance in this extract:

Mr Jones our records show that your last payment was for £500 and it should have been for £750.
My records show that £500 was the right sum.
Well I'm sorry that's the position.
I have the records in front of me and they show clearly that it is £500.
Well we could undertake a further check.
Yes, please would you do that. I'm convinced that my records are correct and that it should be £500. How long would this check take?
A couple of days at the earliest.
I'd like you to do that for me. I think you'll find it is £500.

Mr Jones is holding his ground. He is being assertive but staying calm and not losing his cool. One of the features of assertiveness as we noted on page 54 is that very often you have to repeat your point. Here Mr Jones repeats the point about £500 several times.

Speak clearly.
When telephoning these agencies speak slowly and as clearly as possible. Just think of the difficulties facing those on the receiving end – all those thousands of voices, all those different accents, different forms of English – Newcastle Geordie followed by Belfast brogue with a dash of Glasgow thrown in! Spare them a thought and don't rush into your explanation, or your complaint. Start nice and slowly.

Hello I'm calling to . . . my name is . . . my reference number is . . .

Reach a conclusion.
Be certain at the end of the phone call that you are sure what has been discussed and agreed. You need to have clear answers to these questions:

- what has to be done?
- who has to do it?
- how is it to be done?
- by when has it to be done?

Do take a note of the person you spoke to and any reference number given – the sort that travel operators give you when booking a flight. By using this you will be able to short circuit the complexities of the agencies.

Written communication

Many of your written communications with authorities and agencies will be through forms. Here are a number of typical ones and some hints on how to deal with them. We noted in the section on application forms on pages 43–7 that you need to take great care to fill these up correctly if you are to have much chance of gaining an interview. Well the same is true of forms for agencies and authorities. If you do not respect their wishes when it comes to filling in the boxes then you will have the form sent back and be asked to do it again. So aim to get it right first time.

Planning applications

In many parts of towns, cities and villages there are conservation restrictions. If you carry out any building work you are required to complete a form from the local planning authority. This form is used to gain permission to start any building work. Like all forms they need to be taken carefully. Here's an example.

PLANNING SUBMISSION
The Building Act 1984   The Building Regulations 1991
Elborough Planning 22 Railway St Elborough EB3 4RA

This form is to be filled in by the person who intends to carry out building work or their agent. If the form is unfamiliar please read the notes on the reverse side or consult the office indicated above.

Applicant's details (See note1)
Name...........................   Address......................................
                                   Postcode...................... Tel....................

Agent's details (if applicable)
Name...........................   Address......................................
                                   Postcode...................... Tel....................

Location of building to which work relates
                                   Address......................................
                                   Postcode...................... Tel....................

Proposed work
Description..

Use of building (see note 9)
1) If new building or extension state proposed use............................
2) If existing building, state present use..
3) The premises are/are not intended to be put to a use which is a designated use for the purpose for the Fire Precautions Act 1971 (delete as appropriate)

Fees
1) If category 1 and 2 work state floor area..
2) If category 3 work state number of dwellings...............................
3) If category 4 work state estimated cost of work excluding VAT
    £.........................
        70% of estimated cost of work excluding VAT £.........................
Plan fees £............................   Total £........................

Planning

If proposal has a valid Planning Consent or a Planning Application is currently in progress please give reference no....................................

Statement

This notice is given in relation to the building work as described, and is submitted in accordance with regulation 11(i) and is accompanied by the appropriate fee. I understand that further fees will be payable following the first inspection by the local authority.

Name........................... Signature........................... Date.................

Before filling up this important form you should do what we suggested you do with an application form (page 43). Try it in pencil first. Practise writing it out before you complete it in ink.

Do make use of the supplementary notes to assist you. Where it states see note 1, note 3, note 9, then look these up before you complete the form.

For instance note 1. states that

The applicant is the person on whose behalf the work is being carried out, e.g. the building's owner.

and note 3 deals with the all important issue of fees:

Subject to certain expectations a Full Plans Submission attracts fees payable by the person by whom or on whose behalf the work is to be carried out. Fees are payable in two stages.

The first fee must accompany the deposit of plans and the second fee is payable after the first site inspection of work in progress.

This second fee is a single payment in respect of each individual building, to cover all site visits and consultations which may be necessary until the work is satisfactorily completed.

Remember if all else fails then do phone up the planning office and ask for clarification. They state this at the top of the form but surprisingly few applicants bother to take up this invitation. It is far better to get it clear 'from the horse's mouth' and fill up the form correctly than to have to do it all over again and so delay getting approval.

Such clarification not only saves you time but saves officials work – they don't want to have wasted journeys inspecting your site or sending back half completed forms. So never feel concerned about asking for their assistance. That's their job.

The DVLA (the Driver and Vehicle Licensing Agency Swansea)

If you've ever bought a car, sold one or inherited one from a family member then you'll be familiar with the DVLA. As far as work is concerned you may be required to register a new vehicle or one you have bought second hand. If you deal with a trader he/she will take care of the form filling but if you act privately then you need to read the notes attached to the form.

Activity

Examine the first part of the DVLA form. Ask yourself whether it has been correctly filled in.

B Mr 1 ☐ Mrs 2 ☐ Miss 3 ☐	Please tick the relevant box
....	
....	
Title or business name	*Mr J A Smith (Builder)*
....	
....	
Forenames in full	*John*
....	
....	
Surname/	*Smith*
DVLA Fleet no	
....	
....	
Address	*41 London Rd*
....	
....	
	Elborough
....	
....	
Post Town	
....	
....	
Postcode	*EB4 7TR*

You should have noted the following.

- He has not ticked the box relating to his status. Instead he has written Mr on the next line. There's no great harm in this you may think but these forms are designed to be computer scanned; by failing to tick the box as stated you run the risk of your form being delayed in processing – in other words you may have to wait for your licence.
- Is it clear to the reader of this form – the person in DVLA offices Swansea (where they receive several thousand of these forms per day) – whether Smith (builder) is the business name, i.e. Smiths Builders Ltd, or whether Mr Smith is simply informing the DVLA that he is a builder?
- It asks for forenames in full. He has put J A Smith and yet openly writes John as his forename. What does the A stand for? Albert? Alfonso? Arthur? Or simply Absent Minded? If the form says all forenames – names that comes before your surname – then put them in even if you never use them or wish that your parents had never given you them.
- If the form has a box for post town then do write your town/city in it. These forms are carefully designed so do not leave blanks. If the title of a box is not applicable to you then write NA(not applicable) or put a line through it. If you leave any part of the form blank, as we mentioned on page 43, you give the impression that you have forgotten it.

Application for a passport

Sooner or later you will need a passport. The Passport Agency handles all applications in the UK. You need to go to your local post office and get the appropriate form. State if this is your first application for a passport, or if you are making an application to replace an expired one (usually after 10 years). A different form is needed in each case.

The first point made in any passport application form is:

> Please write in capitals and in ink

So there's no great future in sending an application form in pencil and ordinary writing.

The first question in both forms is:

> Enter date of travel

It is very important that you do this. This helps to ensure that the agency processes your application in time, but remember they will normally need at least one month to achieve this, longer if you apply between January and August. So there's no point in putting up a date only a couple of weeks before that holiday or honeymoon – you'll be disappointed and so will your partner!

Further down the form there's a box in which it asks:

> Daytime telephone number if we need to contact you urgently

It is important that you complete this. If for any reason the agency run into problems with your application – details relating to a parent's place of birth, your grandmother's maiden name, etc. – then very often things can be quickly sorted on the phone and delays kept to a minimum.

As we stated at the start of this chapter all agencies and most of the authorities you will need to deal with in business now have to measure up to targets of performance. One of these targets is the ability to process applications quickly and efficiently. That means making sure that you get the papers you expect to get, within a reasonable time as stated in their blurb. You can help by:

- reading the whole form
- completing all the sections
- using CAPITALS where it tells you to do so
- putting in any necessary enclosures – cheques, postal orders (not cash)
- signing and dating the form.

Conclusion

Communication with authorities is seldom as bad as it seems. It is getting easier since most authorities have made and are making real efforts to reduce the burden of communication. Most now have helplines which enable you to phone up and ask for help. Most have streamlined their procedures and reprinted their forms in plain English. You have to do your part by reading the forms carefully, and any notes that accompany them, and by taking a little time and trouble to get the job done right.

If you do have difficulties then you should certainly phone up or write for assistance. Increasingly these agencies are making us of the WEB so that those who are connected to it are able to get advice on a

24 hours, 365 days a year basis. Whether it's through the WEB, a telephone call, letter or fax, it's in everyone's interest to speed up the communication between the individual and an authority.

Group exercise

In your groups discuss individual experiences you have had with authorities – tax, DVLA, DHSS, NHS, Passport Agency, Job Centres, etc.

(1) Try and list those aspects of the communication that were excellent or good, and those that were disappointing or poor.
(2) Consider what lessons, if any, there are from this for companies in the construction industry wishing to improve communication with their customers.

7

Kit bag

In this 'bag' we've put a number of 'tools' which you will need for your communication. We're not saying that you will need all of these all the time you're at work but we can promise you there'll be no wasted tool. If one isn't immediately applicable at work or college then it will probably come in useful when you you've worked for longer or set up in business on your own. There may also be a use for these tools in outside-work activities and interests – sports clubs, going on holiday, voluntary work, involvement in school/youth groups, etc.

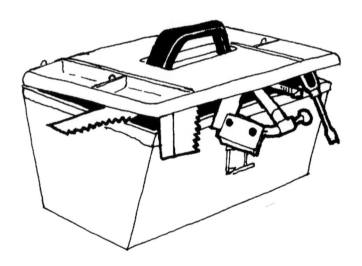

Layouts for letters

We've looked at various types of letters – covering letters to go with application forms (page 47) and letters to customers (page 130). Basically there are two standard layouts.

Use this one for any hand-written letter.

```
                          Your address
Their name
Job title                 Tel no
Their address             Date



Dear Mr/Mrs/Ms/Miss or use first name (if you know
the person)


Reference no/ Job/ contract No, etc. (underlined
or bold


          ........................................
........................................................
........................................................
...........


          ........................................
........................................................
........................................................
...........


          ........................................
........................................................
.........................................................
...........


     Yours sincerely or Yours/Best wishes, etc.


       Signature (print name alongside/underneath)
```

Use this one (called 'blocked') for any word processed letter.

```
              Firm's name on pre-printed stationery

                                            Firm's logo

Your name and title (e.g. Supervisor)
Firm
Firm's address
Telephone/fax/e-mail

Any reference number/contract number
Date

Name of person you're sending letter to
His/her title
Address


Dear Sir/Madam or Mr/Mrs/Ms/Miss or first name

    ..............................................
    ..............................................
    ..............................................
    ..........

    ..............................................
    ..............................................
    ..............................................
    ..........

    ..............................................
    ..............................................
    ..............................................
    ..........

Yours faithfully for Dear Sir/Madam
Yours sincerely for Dear Mr/Mrs/Miss/Ms
Best wishes/Yours  when you just use a name, e.g.
Dear John

Signature (print your name alongside or
underneath)
```

Notice with the blocked version all the lines start at the margin, whereas with the first format you indent the line (by an inch or so) at the start of each paragraph.

Taking notes

You will need a record of what is said at a meeting or on the phone. You may have to take notes from the pages of a log or from a set of instructions, etc. Here are a few ideas for you to think about.

- If you take any notes then don't put them away in your pocket. Write them up quickly or make use of them in some way. After a couple of days it will be impossible to remember what you meant by:

  See H Mon m. 1st- no del cpflr- cd b Thurs?? Phone J wrng chwksched

It did actually mean:

  See Harry first thing Monday morning. Delivery of chipboard flooring delayed until Thursday. Phone Jim to warn him to re-arrange schedule.

You see what we mean!

- Never cramp up your notes. Leave <u>plenty of space</u> between lines. They will be a lot easier to read back to yourself and to make sense of. They'll also make a lot more sense to someone else who may have to read them. Leave wide margins for drawings or someone else's comments.
- Make use of <u>diagrams, drawing, sketches,</u> etc. to help you record what happened and to help someone else get the picture of what's going on.
- Get used to making your own <u>abbreviations</u>. As long as you can read them back and understand what it is they mean then do use them. Most words can be shortened but still be recognisable if you take out the vowels (a, e, i, o, u). e.g:
  —could – cld
  —should – shld
  —requests – rqsts
  —management – mngmt
  Try writing like this – it'll save you a lot of time and effort. (Try wrtng lk ths – it'l sve yu lt tme & efrt.)

Taking notes from a book, document, etc.

You may read something at work which you feel might be useful to you. You don't want to spend money buying the text so it's very useful if you can jot down the key points. This skill of note taking from a text will be useful during your college work or if you continue more advanced studies later in life.

Activity

How would you make quick notes on this information?
Here it is in full. Have a go before looking at our version.

Drainlaying
Fill a 250 mm length of 150 mm pipe standing upright on a flat surface with representative soil sample from the trench. Strike off level with the top. Do not tamp down. Remove pipe and place beside collapsed soil sample. Fill the pipe with a quarter of the material at a time. Ram this with a metal hammer (one with a 36 mm head) until the whole sample has been used. Tamp material until no further compaction takes place – leave surface level. Measure the distance between the level of the spoil and the top of the pipe. If the distance is more than 76 mm the soil is not suitable, if less than 76 mm the material is suitable for different conditions as follows.

Distance – level of spoil/top of pipe	Suitability of material
Up to 50 mm	Suitable for back and side filling in all conditions
Up to 25 mm	For all surrounding pipes jointed to structures where settlement is expected
38–50 mm	For waterlogged conditions
51–75 mm	Suitable only on dry sites, care must be taken on compaction.

Here is one way of noting this information. Are these useful and accurate notes?

```
Drainlaying

Fill 250 mm lgth of 150 mm pipe
Uprght on flat surf with rep soil sample frm trnch
Strike off levl with top xtamp dwn
Remve pipe - place along collpsd soil sample
Fill pipe 1/4 matrl at a time
Ram - metl hammr (36 mm hd) - smple all usd
Tamp matrl - no more compctn
Leave surf levl
Measre dstance - level of spoil - > top of pipe.
If distnce> 76 mm - soil xOK
If <76 mm - soil √OK - see table.


Dist - level spoil/
top of pipe            Materials

> 50 mm                OK back + side filling - all
                       conds
> 25 mm                All sur pipes jntd to
                       strcts - settmnt expetd
38-50 mm               Watloggd
51-75 mm               Dry sts only !! compactg!!
```

You'll see in the above example most of the words have been shortened – distance to distnce, compaction to compctn, material to matrl. Various other short forms have been used:

- < less than, > more than
- x do not; xOK not good, not suitable; √OK = suitable.

Find your own way of doing this. If you can write using these shorthand forms then you will be able to get down a lot more much faster. Do remember not to leave it very long before you translate your shorthand notes into plain English for someone (or for you) to read and understand.

   Some people like mind mapping. That is they put the topic, in this case drainlaying, in the centre of the page and jot down the various points as though they were branches of a tree or spokes of a wheel. You can add the minor points as though they were smaller branches or further spokes. Some people find this a much better way of noting than just writing lines on a sheet of paper. If you have never tried mind mapping perhaps you should. Here's an example from the material on pages 98–111 when we looked at ways of communicating with our customers. We wrote it as a series of lists. Here we could mind map it as follows.

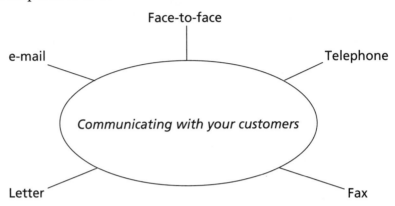

Taking notes from a telephone call

This will be made much easier if you have a telephone message pad. Get into the habit of using such a pad. If there isn't one at work then try to get someone to buy one or you get one and show them just how much more efficient it is to use. If you're in business yourself do use one and don't rely on scraps of paper.

Activity

Examine this way of taking a message from a phone call.

How might it have been better noted down?

Have a go yourself and then compare yours with the version underneath which uses a message pad outline.

On a scrap of paper

Mrs McQueen Leaking roof? water into hall – drips.

Workmen on roof last month fixing new aerial.

Any connection?

Tel 0112–223–345

Worried. Did roof check for her. Don't phone before 1pm.

1st noticed drips Weds night.

Roof check done Nov.

Tel 0112–223–345

Call her ASAP.

On message pad

Telephone Message

Date July 29th 1999

Time 1.30 pm

Caller Mrs McQueen
Phone No (after 1pm) 0112-223-345
Fax/e-mail
Address 21 Ironside Rd Elborough

Message
She's noticed drips of water into hall (from
roof?). Had roof checked last Nov by us. Workmen
up on roof recently installing new aerial. Might
be cause of problem. Said you'd call back ASAP.

Message taken by

Harry Jones

You'll notice that in the first version the person taking the call has just jotted down the points as the caller has spoken them – she's probably a bit upset, even angry (after all you did the roof check for her). The various points of information are scattered all over the place. It would be much more helpful to have the various items of information such as the drips of water, the date of last roof check, the workmen on the roof putting up the new aerial, etc put into a more systematic form.

When there are a number of callers in any one day having such a format for taking notes is very useful. You can imagine the potential for confusion if every phone message is put on a scrap of paper and these scraps of paper get blown on the floor, etc. To succeed in business you need to be efficient. A simple telephone message pad is one of the cheapest and easiest ways to become more efficient.

It's also useful to have a record on the pad of *who* actually took the call – you can go and ask them for clarification of any details – and the *time* the call was taken. It might have been an emergency job.

The other advantage in using a message pad is that messages may get passed on to you over a mobile link. It will be so much easier for you to take notes if someone is delivering information to you in this structured form.

It is also much easier to remember information if it is given to you in a structured way as from a message pad rather than as scraps of information with names, numbers, leaking roofs all jumbled up.

Spelling and punctuation

It's all very well writing messages and letters but what about the nuts and bolts of spelling and punctuation? We're not trying to insult you by giving you basic lessons in writing but you might be a little rusty on certain points or simply might never have been taught them at school. There are many excellent books available which we refer to at the end of this text but here are some key points.

Spelling

One of the problems most people have is with words with the same sound but different meanings. Dictionaries are not very helpful here since you have to know which spelling you need. Spellcheckers on PCs also usually fail to spot the differences. Here are the most common ones. Try to learn and remember the differences. You might like to add your own 'favourites' to this list.

---

advice: My <u>advice</u> is chuck the job now.
advise I <u>advise</u> you to take this college course.

affect The fall <u>affected</u> his confidence.
effect The <u>effect</u> of the fall lasted two months.

compliment He <u>complimented</u> them on their work.
complement The new design will <u>complement</u> the roof structure (i.e. go with it).

counsel He was <u>counselled</u> on his drinking.
council The town <u>council</u> support business.

current Switch off the electric <u>current</u>!
currant I dislike <u>currant</u> cake – give me sponge.

dependant Your elderly mother is your <u>dependant</u>.
dependent His future is <u>dependent</u> on these exams.

devise Did she <u>devise</u> this scheme (i.e. think it up)?
device It's a sensible safety <u>device</u>.

emigrate She <u>emigrated</u> from Spain to France.
immigrate He <u>immigrated</u> into Britain.

ensure   Ensure you've completed all safety checks.
insure   Did you insure against theft?

exhausting   The race was exhausting.
exhaustive   The police search was exhaustive (very thorough).

flare   I noticed it flare up dangerously.
flair   He's got a real flair for the game.

its   The accident log lost its cover some time ago.
it's   It's not important – leave it. (short for it is)

license   I license you to marry this man.
licence   Have you got a full driving licence?

passed   You mean you passed by without stopping?
past   It's all in the past – forget it.

personal   It's personal – don't gossip about it.
personnel   All personnel must wear hard hats.

principle   It's an important principle. (idea, belief)
principal   The Principal of the college. (main one)

practice   She runs a busy doctors' practice.
practise   Great footballers practise their skills.

key   Here's the front door key.
quay   The boat's moored up along this quay.

stationary   The car was stationary.
stationery   Envelopes, files, papers – basic stationery.

they're   They're lost (i.e. they are).
their   Their tools?
there   I put them over there.

too   It's too heavy.
to   Going to the bank?
two   Two pounds of apples.

waste   It's a real waste throwing them out.
waist   Get him by the waist and pull.

you're   You're a damn fool (i.e. you are).
your   It's your fault we've lost.

There are plenty more of these difficult words. But if you can learn these you'll have an advantage over many others and you'll be able to make better use of a dictionary.

Punctuation

Exercise

Each of the following sentences requires some punctuation. Put in the commas, full stops, apostrophes, etc. that you think should be included. Answers below.

1. The customers letter included a cheque for £50 but dont cash it
2. Although we installed the new wiring fuses are still blowing.
3. Mr Fred Smith the general manager of Smiths builders gave a talk on safety.
4. Remember to check the following
    tool box
    hard hat
    ear mufflers
    heavy duty gloves

Answers

Question 1

The customer's (1) letter included a cheque for £50 but don't (2) cash it (3).

(1) We need an apostrophe 's. These show who or what owns something, e.g.

  the worker's tools
  the dentist's drill
  the footballer's boots, etc.

When there is more than one person who owns something, the apostrophe comes after the s, e.g.

  workers' tools (i.e. many workers)
  dentists' drills
  footballers' boots

With people, children, women the apostrophe always comes before the s, e.g.

  people's safety
  children's shoes
  women's clothes

With people's names which end in an s, such as the author's Ellis, you can either write

Ellis's books, or
Ellis' books

There is no need to use apostrophes when writing short forms, e.g.

NVQs not NVQ's

Never use an apostrophe for plurals of words, e.g.

trowel trades not trowel trade's
wood occupations not wood occupation's
joinery crafts not joinery craft's
all paints at half price not all paints'(or paint's) at half price

(2) We do need apostrophes (2) when we leave out a letter, e.g.

don't for do not
can't for cannot
we're for we are
couldn't for could not
shouldn't for should not, etc.
the architect's coming today for the architect is coming today

(3) Full stops. Needed for the end of sentences. They should be
followed by a capital letter.

There is no need to put full stops after initials, e.g.

R.J. Bolton write as R J Bolton
B. B. C. write as BBC

Note Be very careful when writing down e-mail addresses that you only put the full stops (dots) in the right place, e.g.

J.M.Smith@constrct.demon.co.uk

If you place the dots in the wrong position the message won't get sent.
Never put a dot at the end of the address.

Question 2

Although we installed the new wiring, (1) fuses are still blowing.

(1) We need a comma here after an 'although' expression. The same
   would be true after the following.

   Therefore, we have to buy more tools.
   However, she did pass the exam.
   Lastly, he pulled the plug.

There are many other uses of the comma. Use it to provide emphasis
in what you write. You can do this by <u>underlining</u> or, if using a word
processor, by using **bold.** However, a more polite way is to use commas.
   Compare these two statements.

   Please apply, before Thursday, if interested in this post.
   Please apply before Thursday if interested in this post.

By putting commas round 'before Thursday' we are reminding the
reader to hurry up.
   Compare these two statements.

   Please deliver outstanding money in cash before end of the month.
   Please deliver outstanding money, in cash, before end of the month.

The second version has more of a punch, don't you think?
   We also use a comma after

   Dear Mr Smith,

in a hand-written letter. There is no need to use it when word
processing a letter.
   Use it after a so expression, e.g.

   There were few CDs in the shop, so she went into town to buy her
   present.

   Use it to ask questions when writing in an informal style, e.g.

   This is the right tool, isn't it.

   Use it to separate out descriptions, e.g.

   The wall was long, badly pointed and in danger of collapsing (i.e. it
   was very close to being a complete disaster.)

Normally you do not need to place a comma before the final *and* in a list like this. However, if you really want to emphasise the *and* then you can.

The wall was long, badly pointed, and in danger of collapsing.

Use it to separate out parts of dates, addresses, etc.

Wednesday July 14th, 1999 is a public holiday in France.
He lives at 145 High Street, Elborough, Cheshire.

Use it to separate out sets of figures, e.g.

In 1998, 30 new houses were built.

Use it to separate out parts of numbers, e.g.

1,456   1,350,000 (nowadays most printed texts do not use this)

Use it to separate out identical repeated words, e.g.

Whatever is, is right.
Try and tap in, in the hole.

Question 3

Mr Fred Smith the General Manager (1) of Smiths builders gave a talk on safety.

(1) We need capital letters for General Manager as this is a title.
    We need capital letters for beginning a sentence. Use them for: people's names

Smith, Jim, Susan

places

Birmingham, Essex, New York

titles

Mr, Miss, Head Teacher, Dr, Senior Architect, Chief Planner

copyright trades names

Coca Cola

Short forms

BBC, TUC, FA

Question 4

Remember to check the following: (1)

tool box
hard hat
ear mufflers
heavy duty gloves

After the word which introduces the list (in this case *following*) use a colon (:). When you write a list then start each item with a capital letter, e.g.

Tool box
Hard hat

Other aspects of punctuation

The question mark (?)
This is used at the end of a statement put as a question.

Did the architect see the plans?

The exclamation mark (!)
This is used to indicate strong feeling, or humour, e.g.

I'd like that money – how about a Christmas present!
Come back to me when you've won the lottery!

The dash (–)
This is used in informal writing, short notes and friendly hand-written letters, e.g.

I hope this reaches you on time – if not give me a phone.
The architect's coming on Tuesday 11 am – could you be here then?

The hyphen (a dash between parts of words)

There are some words that always need to be hyphenated, e.g.

re-cover (your sofa) compared with recover (getting better from illness)

You should also use a hyphen with such expression as off-putting, in-situ (in place). There are hundreds of words where you can use hyphens if you want, e.g.

co-operation and co-contractors (it prevents the double oo)
mis-spelling (preventing the double ss)
sub-contractor compared with sub contractor

There is a also a need for a hyphen where a word is used with a specialised meaning, e.g.

In the space below draw a view of a hop-up.
Just hop up on the lorry and get some bricks.

Writing clear sentences

The sentence is the basic form of writing. It should make sense by itself.

Activity

Are these sentences?
 On the line
 By the way
 You will get

The answer is NO. They do not make sense by themselves. We can make them into sentences by adding further words.

Is it on the line?
By the way have you got the money?
You will get paid when I sign the contract.

By adding these words we turn phrases into sentences which make sense by themselves.
  Here are two points to watch when you're writing sentences. (Notice the apostrophe in you're – short for you are!)

Don't make them too long. (Notice the apostrophe in don't!)

The longer the sentence the more difficult it will be for most of your readers to understand.

Activity

How would you make this sentence shorter and easier to read?

There are three things a fire 'needs' to keep burning: fuel, oxygen and heat, if the fuel – anything that can catch alight – can be removed from the area of the fire, this will reduce the potential damage that will be suffered.

It could be re-written as:

There are three things that a fire 'needs' to keep burning: fuel, oxygen and heat. If the fuel – anything that can catch alight – can be removed from the area of the fire, this will reduce the potential damage that will be suffered.

These two sentences of 15 and 27 words will be easier to read than one long sentence of 42, especially if your reader doesn't have a lot of time for browsing.

Remember that many long sentences which go into that kind of detail can be written as lists, e.g.

There are three things a fire 'needs' to keep burning:
—fuel
—oxygen
—heat

Activity

Here is another example of a long sentence. How would you chop it up for improved ease of reading?

The storage and stacking of roofing materials is very important, for instance check that rolls of felt are always stood on their ends on a clean boarded or concrete standing and the bitumen neatly piled on a similar clean platform.

You could break this long sentence into the following three sentences:

The storage and stacking of roofing materials is very important. For instance, check that rolls of felt are always stood on their ends on a clean boarded or concrete standing. Check also that the bitumen is neatly piled on a similar clean platform.

Make sure that your sentences are clear for your reader.

Ensure that there is no ambiguity – i.e. more than one meaning. It's no good writing sentences that only you can understand. The trick is can *your readers* understand them? For example it may be funny to read these sentences in joke books:

Monster Sale.
Dogs. Please shut these gates.
Full and half cooked breakfasts served.
Several constantly rotating guest beers served all day.
He had been driving for 30 years when he fell asleep at the wheel
Stranger battered to death in fish shop horror.
Dear dentist, the teeth in the top are OK but the ones in my bottom are hurting badly.

But we must be careful – unless we are trying to be funny – not to let our writing become ambiguous.

Activity

Read this sentence carefully. Is it clear to you what the writer intends?

Wiring will be carried out on the looping in principle, all such joints being made at main switches only, and at sealing boxes, sockets, lighting outlets and switch boxes.

Does it mean that the joints will be made at the main switches and also at the sealing boxes, sockets, lighting outlets and switch boxes? Or does it mean something else? What then is the point of writing *only*? Would it have been better to have written:

Wiring will be carried out on the looping in principle, all such joints being made at main switches and at sealing boxes, sockets, lighting outlets and switch boxes only.

By putting only at the end of the sentence does that make the meaning clearer? How about this version?

Wiring will be carried out on the looping in principle, all such joints being made only at main switches and at sealing boxes, sockets, lighting outlets and switch boxes.

Would it have been better to have added a comma before the and to give it even more emphasis?

, and at sealing boxes, sockets, lighting outlets and switch boxes

As well as being careful with the word *only*, you must take care with *this* and *it*. If you name two or more items then what does the *this* or the *it* mean to your reader? e.g.

Before glass is placed in the rebate this must be back puttied.

In this sentence the meaning of *this* is fairly obvious but there is just a bit of doubt.

Activity

How would you re-write it so as to cut out any doubts?

How about

Before the glass is put into the rebate it must be back puttied.

Does that sort out the problem? Will the reader be clear as to what the *it* refers to? How about

Before placing in the rebate, the glass must be back puttied.

Is this clear? Will there be any doubts now? Finally,

The glass must be back puttied before it is placed in the rebate.

That reads more smoothly and cuts out any doubts as to meanings.

Activity

Have a look at this sentence.
Is it clear enough? Would you change it?

All wall tillings should be finished with straight and level joints, free edges, level or plumb as applicable, they should be protected against irregularities or distortion.

Did you think that meaning of the *they* was clear enough for the reader? Is it 100% obvious that it is the wall tillings that should be protected?

There's a great deal more we could say about ambiguity in writing. Keep your eyes open and try to spot those efforts which are ambiguous – some of them might make you laugh.

Communicating through graphics

A great deal of information that we read comes in the form of tables, graphs, diagrams, flow charts, etc. It is helpful to know something about these devices. Graphics can be highly misleading. Which of these lines is longer?

Most people think that the bottom one is, but this is just an illusion. So we need to be on our guard when we look at tables, charts, graphs, etc. Here is a quick run through of the key forms.

Pie charts

The pie chart – named because the various bits of information are shaped like a pie – has strong visual appeal. You can quickly see the way the various bits of information contrast with each other. For example, suppose we have the following information concerning the proportions of different types of work done by a building company.

- New homes 40%
- Restoration work with older properties 15%
- Contract maintenance work of properties 30%
- Other work 15%

We can show this as

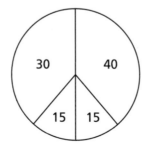

Pie charts are very useful when the various numbers are large – 10% and bigger. They are much less satisfactory when you have a whole lot of small segments as in this chart which shows the work of Smiths Builders 1990–1997.

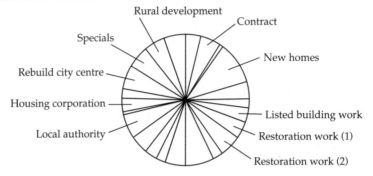

This looks messy and it would be better to find another way of showing the information.

Tables

The whole point of a table is to provide the reader with information at a glance. For instance here are fatality figures in service and manufacturing sectors for 1997/8.

Activity

Examine this table very closely. Where do you think it should be improved.

<u>Fatality Figures</u>
The number of fatal injuries reported to all
enforcing authorities by industry sector
are estimated for the year 1997/98 as follows:

EMPLOYEES
Service industries   64
Construction   54
Manufacturing   54
Agriculture, forestry and fishing   21
Extractive and utility supply   17
(incl. offshore oil and gas figures)

SELF EMPLOYED
Service industries   10
Construction   22

Manufacturing   7
Agriculture, forestry and fishing   19
(ex sea fishing)
Extractive and utility supply   0

You might have spotted the following problems with this table. The underlined words and phrases are those we suggest should be questioned.

Fatality Figures
The number of fatal injuries reported   *Should we know what this*
to all enforcing authorities by industry   *means?*
sector are estimated for the year   *Should we have details on*
1997/98 as follows: *this?*

EMPLOYEES
Service industries   64
Construction   54
Manufacturing   54
Agriculture, forestry and fishing   21   *Does this exclude sea fishing?*
Extractive and utility supply   17
(incl. offshore oil and gas figures)

SELF EMPLOYED
Service industries   10
Construction   22   *Should we try and line up the*
figures to make them easier to see?

Manufacturing   7
Agriculture, forestry and fishing   19
(sea fishing)   *This does exclude sea fishing?*
Extractive and utility supply   0   *Does this include offshore and*
gas figures as above?

Here are two important points to note when constructing tables.

Make sure that the various columns line up under each other.
We don't want any of this.

Number of employee Smith Builders	1995	1996	1997	1998	1999
Total employed		42	39	38	32	33
males	37	31		30	29	27
females	5	8	8	3	5

Take a little more time and arrange your tables so that your reader has no difficulty in picking up the relevant information.

If you have use of a computer click the insert file and in most programs you will come across *tables*. When you have clicked on this you will be asked how many rows and how many columns you require for your table. The beauty of this program is that it will produce a very neat table with no straggling numbers and uneven margins. e.g.

Smith Builders

Employees	1995	1996	1997	1998	1999
Total	42	39	38	32	33
Males	37	31	30	29	27
Females	5	8	8	3	5

Avoid making your table too long. The longer the line of information the more likely it is that your reader's eye will slip. For instance avoid this.

Smith Builders

Year	82	83	84	85	86	87	88	89	90	91	92	93	94	95
Staff No	45	47	48	51	50	52	50	55	56	56	49	56	55	56
Full/T	22	23	22	21	24	25	22	20	24	26	24	25	26	27
Pt/t	20	20	22	23	22	21	24	25	27	22	22	25	24	20
Other	5	4	4	7	4	6	6	5	5	10	3	6	5	9

Activity

How would you alter this table so as to make it easier to read at a glance?

One way would be to re-order the table so that the year is placed on the right hand side and the different staff types placed above. This will result in a stubbier table. The lines will be shorter and so easier to read.

Smith Builders

Year	Staff No	Full/T	Part/T	Other
1982	45	22	20	5
83	47	23	20	4
84	48	22	22	4
85	51	21	23	7
86	50	24	22	4

87	52	25	21	6
88	50	22	24	6
89	55	20	25	5
90	56	24	27	5
91	56	26	22	10

92	49	24	22	3
93	56	25	25	6
94	55	26	24	5
95	56	27	20	9

Notice how after every five columns a blank space is introduced. This helps to break up the figures which again makes it easier on the eye in terms of absorbing the information.

Bar charts

These consist of rectangles or straight lines. We could display the information using bar charts. It is helpful for the reader if you place the number at the top of the individual column.

Numbers of staff 1999	Full-time	Part-time	Casual

```
-60

           ┌─ 53 ─┐
-50        │      │
           │      │
-40        │      │     ┌─ 37 ─┐
           │      │     │      │
-30        │      │     │      │     ┌─ 23 ─┐
           │      │     │      │     │      │
-20        │      │     │      │     │      │
```

Flow charts

These show how systems operate – procedures. We saw a very simple one on page 66. This depicted how cement was mixed. Flow charts can be that simple or very complex depending on the nature of the process being described.

The Gantt chart

This is used when you want to make comparisons between real achievements and planned performance, e.g. number of houses planned for building in a one year period as against actual completions.

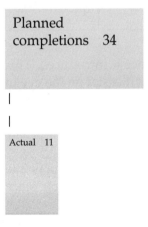

Planned
completions   34

Actual   11

Line graphs

These show a dynamic relationship between two functions. The Y is plotted on the vertical and the X on the horizontal. As an example we could plot the relationship between a steady rise in temperature (Y) and the expansion of a joint in a metal frame (X).

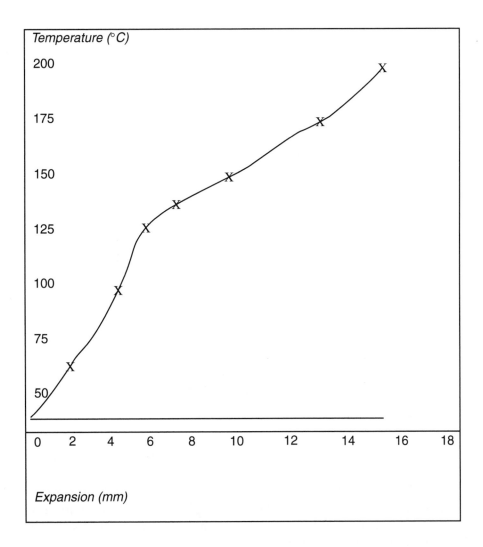

When you are constructing a graph it is a good idea to try it out with pencil, ruler and paper before you do it in ink or on a computer. Draw the various plot lines between X and Y. Try out the various scales since there's no point in having the scale too small so that the reader cannot see what it is you're trying to plot.

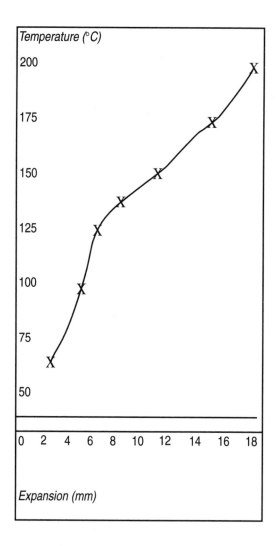

This scale is may be just too small to be of much use to your reader.

Another point: although it can be very interesting and revealing to place more than one set of data on a graph be careful that you don't confuse your reader by overloading him or her with too much information. Keep it simple.

Useful short abbreviations and short forms

There are so many of these flying around these days that it is almost impossible to keep track of them.

Exercise

See how many of these you can provide the appropriate abbreviation for.
Answers on page 193.

metre	hectare
square metre	kilo
watt	ampere
litre	kilogram
volt	degree Celsius
centimetre	hundredweight
millimetre	

Exercise

See how many of these abbreviations you can write the full name against.
Answers on page 193.

DVLC	PAYE
c o d	PTO
misc	s a e
mpg	M
VAT	t

Please note the following points.

- When using a short form for the first time always write it in full if there is even the smallest chance of confusion.
- Do not put a full stop after a symbol, except when this marks the end of a sentence.
- Do not confuse capital and lower case letters (m for metre and M for mega).
- Do not use hyphens, i.e. millilitre not milli-litre.
- Do put an 0 before the decimal point when the quantity is less than one, i.e. not .8 but 0.8.
- Do not use commas to mark off thousands. Use a space instead, i.e. not 1,000,000 but 1 000 000.
- Do avoid any mixing of units, e.g. 17.75 Kg not 17 Kg 75 g.

Using a library and reference books/computer access

Libraries are often confusing places. You enter them and are surrounded with shelves of books. There's the odd number printed above the stacks of books and perhaps the odd sign such as BUILDING or POPULAR MUSIC. Here are a few helpful hints.

Most libraries use the Dewey Decimal Classification (DDC). This divides knowledge into ten sections:

000–099 General works	500–599 Natural Science and Maths
100–199 Philosophy and Psychology	600–699 Technology/Applied Sc
	700–799 The Arts
200–288 Religion	800–899 Literature
300–399 Social Science	900–999 Geography, History
400–499 Language	

As an example, if you wanted to do some work to improve the performance of your car you would want to look up information relating to the internal combustion engine and how to tune it. Here's how you would search.

600–699 This will give you your start – Technology
   620 would be the next sub division – Engineering
   We're getting closer.
      621 this now gives you Applied Physics
         621.4 gives you Heat Engineering
            621. 43 gives you Internal Combustion Engines
            (We are now getting very close.)

When the Dewey system was first devised in 1876 all the material was contained in 24 pages. A recent edition ran to over 4000 pages with thousands of sub-divisions.

You may find that the library you use does not use this particular system but that devised for the Library of Congress, Washington USA. Many public libraries make use of this system. It has 26 divisions and these are shown by an initial letter. A second letter indicates a sub-division; then numbers follow the letters for further subdivisions. e.g.

A Encyclopaedias and reference books	M Music
B Philosophy, Psychology, Religion	N Fine Arts
C Antiquities, Biography	P Language and Literature
D History	Q Science
E–F American History	R Medicine

G Geography
H Social Sciences, Economics,
   Sociology
I Political Science
L Education

S Agriculture, Veterinary Science
T Technology
U Military Science
V Naval Science
Z Books and Libraries

Some specialised libraries will have their own system but whatever the actual system, catalogues are the keys to any library: they allow you to open up the system and locate your book. Increasingly card catalogues are being replaced by computer-based ones. These allow you enter

the name of the book – How To Tune Petrol Engines
the author/s   – Smith J and H Alexander
the publisher   – Magic Motoring Publishers

If you don't have any of this information then you can enter the subject. It will help your search considerably if you can supply more information than simply:

Petrol engines

The computer program will take you through various questions so that it can locate what you want.

Much useful information can be found in the reference collection. The material in this section is for looking at and reading in the library

but not for borrowing. Here you will find encyclopaedias, handbooks, dictionaries, British Standards, atlases, etc. You will also find material relating to college courses, training schemes, etc.

The most important source for finding information in any library, however, is the human one. Always remember that librarians and their assistants are there to help you. Ask if you are uncertain about where to find something, a book, a CD, a pamphlet, a directory or an address for an organisation. The more information you can supply the easier it will be for the librarian to assist you. If you go to the counter and say:

I'm looking for something on engines.

the librarian will need to ask you quite a few questions. If you can say:

I'm looking for something on fine tuning a petrol engine.

then the librarian has more of a chance of showing you where to look on the shelves. If you can say:

I'm looking for something on fine tuning a petrol engine, in particular sports cars, pre-1970s.

then you really can be helped and speedily, even if it's only a question of *Well we don't have anything like that in stock*. That's a help? Yes it certainly is. There's no point in trying to locate something if it's just not in the library. This is where inter-library loan can help (where one library borrows from another) or where your librarian can suggest from his or her knowledge of other libraries (public ones, specialist ones) where you can complete your search. Providing extra information does help you in any search.

Handling exams and course work

Exams

You may be taking exams as you read this, or you may have passed some and are looking forward to getting the rest out of the way. On the other hand you may not have taken any exams or completed any course work for a number of years and you may feel a little on the rusty side. There are a lot of falsehoods and popular myths about exams. For instance:

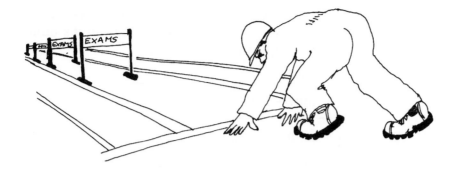

Failing an exam is the end of the world.

No, in most cases there are opportunities for re-sits; you can have another go like the driving test. There are many occasions where although you have actually failed the exam you may still pass the course because you have done well in course tests and practical work and have a good report from a placement.

Exams are set out to trip you up, find your weaknesses, expose you to ridicule and make you appear a damn fool.

No, everyone wants you to do well – including the examiners. They actively look for material in your papers they can give ticks and awards marks to.

You can only pass the exam if you have covered all the ground, been to all the classes and read all the notes.

No, don't worry about what you haven't done, the notes you haven't read; concentrate on what you have understood; work out how to make the very best of what you have done. Obviously if there are large gaps, these will be exposed in the exam.

Exam papers are unreadable, designed to baffle the reader.

Every exam question will be linked in some way with work from the course. Take a few minutes to read through the questions and work out the links. Don't panic!

Exams are fine for those that work fast – the speedsters.

No. Many of those who speed through the questions are in fact answering the questions they want to write about and not those as set by the examiners! It's usually much better to read the questions and then steadily work your way through them. If you can it's quite a good idea to jot down a simple plan:

intro – what to start with
point 1 _____   point 2_____   point 3 _____
conclusion

Exams are OK for those who can spell.

Examiners realise that you are writing under pressure. Do your best to spell words correctly but don't spend minutes working out if it's *necassary* or *neseccary* or *nesessary*; put one down (hopefully *necessary*!) or think of an alternative, e.g. essential, vital, etc.

Having dispelled some of the myths here are some practical suggestions for taking examinations.

Read through the whole paper. Read carefully the instructions and information for candidates printed on the top.
e.g.

All questions must be answered in the space underneath.
Attempt all questions.
Write in blue or black ink.
All questions worth five marks.

So many candidates fail their exams before they've even put pen to paper because they really haven't a clue what it is they have been instructed to do. It is particularly important that you make careful note of the marks and time allotted to each question.

Answer the question that you've been asked, and not the one that you'd like to have been or the one that you thought they'd asked.
e.g.

State FOUR site administrative practices performed by the site manager.

In this case it says four so supply four and not three or five. You might have preferred that it asked you to state the health and safety measures performed by the site manager since that was the topic you had revised but if you do it your way you will only score perhaps 25% of the possible marks. It states administrative practices so answer the question that way.

If you get stuck answering a question, leave it and move on.
You may be able to return later to that question and gain a few more marks by doing a bit more. There is absolutely no point in hammering away at a question, biting the end of your pen, scratching your head and getting over stressed and anxious with the precious minutes ticking away. You would be much better advised to put your pen down, look round for another question and get started on it.

Don't spend too much time on one question.
You may know everything there is to know on this particular topic but if there are only five marks allotted to each question then even if you do a brilliant job on one question you cannot score more than five – there are no bonus points for a really outstanding answer. So do what you can, as best as you can and then move on to the next question.

Use diagrams to help you answer a question.
They enable an examiner to see more clearly what you're trying to do. Don't worry if under pressure your drawing skills fail you; even an approximation will help the examiner see if you've grasped the main ideas. Do label any drawing and indicate key features such as directions (North–South), scale (1 cm = 1 metre), angle of view (side elevation).

A word of warning: don't be tempted to make your diagrams 'pretty' – they're supposed to be sketches not completed architect's plans in full colour. You won't get any more marks for your artistry and you may well over run which means you'll be left with less time to complete the other questions.

Don't cross out work unless you're sure it's wrong.
If you do then it won't be marked and you might be depriving yourself of a couple of valuable marks.

Don't stop working/revising until the time is up.
You never know. Those extra words that you manage to write down might just be enough to take you over the pass mark.

Try and leave a few minutes at the end for checking.
Leave some time to re-read your work, check any calculations, clarify any diagrams and see if you can get rid of any obvious mistakes with spelling and punctuation.

Final point: do write the number of the question you are answering at the top of the page.

Course work and tests

By this we include all those on-site tests which you may be required to do, and the collection of materials which you need to be able to show to an assessor. Here is a checklist for you.

Make sure you know what it is you're supposed to be doing.

> The candidate will be required to demonstrate knowledge and understanding of the need for and methods of identifying short-comings in service to customers.

It should be pointed out to you what *required to demonstrate knowledge and understanding* actually means. If you are not clear then you must speak up and ask. Very often these phrases mean that you will have to collect material and this could be:

* examples of service standards
* examples of standards that you have used
* examples of gaps between practice and the standards as set down.

Here's another example:

The assessor must be satisfied that the candidate can practically demonstrate competence under working conditions . . .

You will need to understand very clearly what is meant by *practically* and by *demonstrate* and especially *competence*. You must not be vague about such words. You will certainly not do well – even fail – the coursework if you do not have a clear understanding of such terms. Ask your tutors. Don't walk away unclear. Ask again. See if you can read something from the City & Guilds NVQ/SVQ.

Make sure you know the deadlines for submission of work or the dates on which you will be assessed.

Make sure that you are familiar with the format in which the course work has to be presented.
By format we mean the way the pages have to be laid out, the headings required, what kind of drawings are expected, to what scale, etc.

Better reading

Having mentioned exams and course work we should briefly look at reading. Many people in the construction industry have to read – reports, health and safety materials, letters from customers, etc. – and many of them also find reading very difficult. Here are some reasons that we've collected from talking to students:

> Boring! Books, etc are basically boring.
> Nothing in most of the stuff at work that's of interest to me.
> Always rushed for time. Never have a chance to sit and read.
> Can't understand the language – too complicated by half.
> Print's too damn small.
> Usually have gloves on, can't hold the paper.
> Books are difficult to get hold of and expensive to buy.
> Not enough pictures, diagrams, etc.
> The only way I can read is on a Spanish beach with a drink in my hand!

Here are some tips to assist you with reading.

Know why you're reading the report, article, etc.
Ask yourself:
• What am I reading this for?
• What's in this for me?
• What can I get out of this?

We're sure you've had the experience of trying to read something and having to stop after every few lines thinking Why Why Why am I reading this? If you can try and find some reason for reading then your concentration on the task will be that much greater and you'll remember more of what you read. The reasons for reading may be varied but could include:

• I should know this for my safety;
• I should understand this because I can do my job better;
• This might save me time and money;

- This could be useful to a mate, member of my family, etc;
- This could help me get a better job, etc.

The question is how do you know if the text/book, etc. is going to be useful? Well you could:

- look at the title or the heading – this should offer you a few clues;
- glance down the contents list if there's one – more clues;
- glance at the blurb – the back of the book – more clues;
- dip into the text, run your eyes down various pages.

In other words get a reasonably good idea of what it's all about – enough for you to know whether you should be reading it or not. Here are a few suggestions for assisting you with your reading (and we don't mean on a Spanish beach with that good thriller!).

Have a pencil in your hand as you read.
It will help your concentration a great deal if as you read you jot down at few notes (see page 155.) If it's your book, or your copy of the report or letter, underline or circle the key words as you take your eyes across the page. Passive reading – just letting the eyes wander along the lines – needs to be replaced by active reading – getting to grips with what the text means. Such underlining and circling of words and lines can help; so can making notes and writing questions in the margin.

Get comfortable.
Sitting in an uncomfortable chair can make the whole process of reading much more difficult. Make some adjustments, sit on something soft and put it behind your back. (Not the Site Foreman!)

Get the lighting right.
Most site offices have a strip light in the middle of the ceiling. This means that very often the light is being obscured by one's head and shoulders. The best kind of lighting for reading is to have the light source falling over your shoulder since it reduces glare.

Check your eyesight.
We don't want to be rude but have you had your eyes tested recently, especially if you are a mature reader? Your eyes can deteriorate without you being aware – the muscles work harder to compensate for any deficiency in vision. Even if you wear glasses for reading then the lenses can become unsuitable. Pride often prevents people from going to have their eye tested but if you find yourself pulling back

from a text or from a noticeboard to see it more clearly then you probably need reading glasses. There are health and safety issues here – obviously – if you can't see to read notices and instructions then you are putting yourself and others at risk.

Thinking about training

Now that we've come to the end of this particular book, which has been about developing your communication skills, it's a good opportunity to take a wider look at training. There are many books around on this subject but here are a few pointers for you to consider.

Assuming you've passed some exams, you may think that's it, you've got your certificate, that's you and training finished. But think about it. You may need to:

• gain more skills if you want to progress to better paid jobs;
• update your skills if you want to stay in your present work;
• learn more skills if you are out of work and can't find much except the odd bit of casual labour;
• retrain if you want to change jobs, move into an new area, start up in business on your own.

Activity

Consider for a moment what prevents people from going for training.
Jot down as many reasons as you can.

Here are some we've gathered in talking to people in the construction industry. How do they measure up to your list?

This job's OK. I don't need to bother.

Well that's being optimistic. Jobs don't last for ever. Security is a thing of the past. Businesses go out of business. You could be left high and dry. New skills and recent training would help secure you another job.

Training's a waste of time. Bosses want to know if you can do the job not if you have bits of paper.

Well that's partly true for the smaller firms but the larger ones increasingly want to see what certificates you've gained and when you gained them (and at what level). Certainly if you want to be able to get work in Europe or overseas then you'll need certificates to prove that you've reached a particular level of skill (as well as good references from recent work here in the UK).

> I'm not clever enough. I've reached my limit with what I did at college – that was difficult enough.

Who says? Most people have a very poor view of their own abilities. Just because you may not have done very well at school or struggled at college doesn't mean to say that you're stupid and won't be able to take any more training. The secret to successful learning is gaining the confidence that you can succeed and the belief that you really want to do it.

> I haven't got the time for study. I'm too knackered after work to go to college in the evenings.

This view is very understandable. However, increasingly colleges are offering part-time courses, there are all kinds of new training opportunities available which are designed to make it easier for you to take part. Many more forward looking employers will consider allowing you time off work for study if this is leading to a nationally recognised award.

Here are some possible sources of information on training.

- Your local college of further education.
- Training Access Points (TAPS). These are computer terminals on which you can find out about training opportunities in your area. TAP points are now increasingly found in local libraries and supermarkets.
- Training and Enterprise Councils (TECs) or local enterprise companies (LECs) in Scotland. They are responsible for ensuring that you get advice and guidance on:
  —training for a new job
  —adding to your existing skills
  —setting up in business on our own
  —expanding your current business
- If you belong to a trade union then it will most probably run a range of courses. You need to get information from full time officials.
- Job Centres. They will have a wide range of information for you relating to training opportunities. They will also be able to inform

you of what grants may be available to help pay fees, travel expenses etc, if you should undertake a course of training.

- The Open College. This has been set up by the government to provide a range of courses (see Appendix 2 for address). It makes use of workbooks, videos, audio tapes, TV programmes and support from local tutors.

Here are two key training terms worth knowing.

Competency-based training

This refers to the way in which qualifications have increasingly been broken down into a range of competencies. They describe in as exact a way as possible what the person is capable of doing. All NVQs (National Vocational Qualifications) or SVQs (Scottish Vocational Qualifications) are based on standards of competence for specific jobs. For example *Building Crafts. Unit 53 Contribute to the Selection and Use of Access Equipment NVQ level 1* sets out a number of objectives. It states that on completion of this unit that the students will be able amongst other things to:

- describe methods of material and component storage;
- describe methods of clearing and disposal of waste;
- list procedures to be followed in response to emergency situations.

These are competencies which gradually build up in terms of level and difficulty as students progress through levels 1, 2, 3, 4, etc.

Accreditation of Prior Learning (APL)

This gives you a qualification for what you already know and can do. If you have worked on sites for a number of years this experience can be brought to an APL adviser at your local FE college and he or she will help you to gather evidence to prove your competence. The evidence will need to be presented for assessment. You can't simply state that you have done this or that; you need to be able to prove it. This might involve you in taking a skills test, gaining references as to the quality and level of your work from employers, collecting photographs of work completed by you, writing up logs of what you've done but in a logical and systematic way, etc.

We've now taken you though this particular training journey. We hope it's been useful. Here are a few final tips.

Observe others communicate.
One of the best ways to improve any skill is by watching others who are 'performing' and learn how they do it. By close observation of others' communication you can learn what to avoid doing and what to adapt for yourself. Remember you can never copy someone else but you can certainly adapt the good 'bits' to your own communication style.

Practise and try and get some feedback.
It's no good practising without gaining any feedback from others. If you have been working through this book with others, hopefully they have given you some feedback. If you never ask others for their opinion as to how you're communicating then you'll never know!

Try and read some of the books we've recommended. (See page 194)

Small improvements can have big effects.
This is especially true when you are at work in your own business. Think about some of the ideas we've put forward which could streamline your communications.

Standards in communication will pay dividends.
This is especially true when dealing with customers. Think how you individually can improve the standards of communication where you work.

Good luck!

A reminder of those 10 questions that you should ask about any communication.

- Is this communication necessary?
- Is it targeted?
- Is it timely?
- Is it in the right language?
- Is it clear?
- Is it accurate?
- Is it short and to the point?
- Does it cover what it should cover?
- Can we get a response?
- Is the tone OK?

Answers to questions

Page 179

m	ha
m_2	k
w	A
l	kg
V	°C
cm	cwt
mm	

Page 179

Driver Vehicle Licensing Centre
cash on delivery
miscellaneous
miles per gallon
Value Added Tax
Pay as You Earn (tax taken off wages)
Please Turn Over
stamped addressed envelope
Mega as in mega watt (a million watts)
tonne
for example

Follow-up reading

There are hundreds of books on communication skills. Here are just a few which will be useful as a follow up to what you've been reading here.

Communication skills

Kelcher M (1992) *Better Communication Skills for Work*. BBC Publications.
> A well set out short book, easy to read, full of useful tips on writing reports, taking an active part in meetings, giving a good presentation and getting the most out of training.

Ellis R and McClintock A (1994) *If You Take My Meaning*. Arnold.
> Practical exercises and an easy to follow guide to key ideas in communication.

Burton G and Dimbleby R (1996) *Between Ourselves*. Arnold.
> A very helpful introduction to interpersonal communication. Particularly strong on transactional analysis.

Writing

Hilton C and Hyder M (1992) *Getting to Grips with Writing*. Letts Educational.
> A very well set out guide to writing, full of advice on the layout of letters and reports.

Cooper B (1987) *Writing Technical Reports*. Penguin Books.
> Sets out key ideas on report writing very clearly; has useful examples of various report layouts.

Appendices

Appendix 1 Your communication checklist

Having read the book here is your chance to assess your skills in order
that you can improve them.

	Excellent	Good	Occasional difficulties	Real problems
Writing				
Letters				
Reports				
Listening				
Speaking				
Talking to people				
Interviews				
Talking to groups at meetings				
Talking over the phone				

Appendix 2 Where you can find more

The Open College
Freepost P O Box 35 Abingdon OX14 3BR
www.opencollege.co.uk

Index